Springer Theses

Recognizing Outstanding Ph.D. Research

Aims and Scope

The series "Springer Theses" brings together a selection of the very best Ph.D. theses from around the world and across the physical sciences. Nominated and endorsed by two recognized specialists, each published volume has been selected for its scientific excellence and the high impact of its contents for the pertinent field of research. For greater accessibility to non-specialists, the published versions include an extended introduction, as well as a foreword by the student's supervisor explaining the special relevance of the work for the field. As a whole, the series will provide a valuable resource both for newcomers to the research fields described, and for other scientists seeking detailed background information on special questions. Finally, it provides an accredited documentation of the valuable contributions made by today's younger generation of scientists.

Theses are accepted into the series by invited nomination only and must fulfill all of the following criteria

- They must be written in good English.
- The topic should fall within the confines of Chemistry, Physics, Earth Sciences, Engineering and related interdisciplinary fields such as Materials, Nanoscience, Chemical Engineering, Complex Systems and Biophysics.
- The work reported in the thesis must represent a significant scientific advance.
- If the thesis includes previously published material, permission to reproduce this must be gained from the respective copyright holder.
- They must have been examined and passed during the 12 months prior to nomination.
- Each thesis should include a foreword by the supervisor outlining the significance of its content.
- The theses should have a clearly defined structure including an introduction accessible to scientists not expert in that particular field.

More information about this series at http://www.springer.com/series/8790

Jordan A. Hachtel

The Nanoscale Optical Properties of Complex Nanostructures

Doctoral Thesis accepted by
Vanderbilt University, Nashville, Tennessee, USA

Jordan A. Hachtel
Oak Ridge National Laboratory
Oak Ridge, TN, USA

ISSN 2190-5053 ISSN 2190-5061 (electronic)
Springer Theses
ISBN 978-3-319-70258-2 ISBN 978-3-319-70259-9 (eBook)
https://doi.org/10.1007/978-3-319-70259-9

Library of Congress Control Number: 2017957647

© Springer International Publishing AG 2018
This work is subject to copyright. All rights are reserved by the Publisher, whether the whole or part of the material is concerned, specifically the rights of translation, reprinting, reuse of illustrations, recitation, broadcasting, reproduction on microfilms or in any other physical way, and transmission or information storage and retrieval, electronic adaptation, computer software, or by similar or dissimilar methodology now known or hereafter developed.
The use of general descriptive names, registered names, trademarks, service marks, etc. in this publication does not imply, even in the absence of a specific statement, that such names are exempt from the relevant protective laws and regulations and therefore free for general use.
The publisher, the authors and the editors are safe to assume that the advice and information in this book are believed to be true and accurate at the date of publication. Neither the publisher nor the authors or the editors give a warranty, express or implied, with respect to the material contained herein or for any errors or omissions that may have been made. The publisher remains neutral with regard to jurisdictional claims in published maps and institutional affiliations.

Printed on acid-free paper

This Springer imprint is published by Springer Nature
The registered company is Springer International Publishing AG
The registered company address is: Gewerbestrasse 11, 6330 Cham, Switzerland

for Helen

Supervisor's Foreword

Jordan A. Hachtel's Ph.D. dissertation was the product of an extraordinary journey of search and discovery. In order to understand Jordan's scientific contributions, one must follow him in this journey. After taking the core graduate courses, Jordan was uncertain about whether to pursue research for a doctorate in theory or in experiments. He came and saw me and I offered him an option to do both. Shortly after I joined Vanderbilt in 1994, I accepted a secondary appointment at the nearby Oak Ridge National Laboratory (ORNL) and spent a day each week there, building a collaborative theory/microscopy program with the microscopy group headed by Steve Pennycook. I already had several graduate students who completed Ph.D. dissertations in theory and microscopy with flying colors. Jordan jumped on the idea.

Initial funding for Jordan came from a joint grant with experimentalist researchers at the University of Tennessee campus in Knoxville on the development of novel photovoltaics. Jordan learned a great deal about making superlattice samples to enhance lattice absorption and quickly got his training in scanning transmission electron microscopy. At the same time, he focused on the theoretical description of light absorption by a superlattice of very thin layers and possible enhancement of light absorption by the unique bonds at the interfaces compared with absorption by a model superposition of bulk-like films. Jordan developed a method to combine density-functional-theory (DFT) calculations with macroscopic methodologies to extract the interface effect on absorption in a consistent and quantifiable manner. He used $NiSi_2$/Si heterostructures and demonstrated that, by varying the relative volume fractions of interface and bulk, one can tune the spectral range of the heterostructure absorption.

As Jordan was finishing the theoretical work, microscopy beckoned with a unique opportunity. Postdoc A. Mouti at ORNL had built a cathodoluminescence (CL) system in a scanning transmission electron microscope (STEM) and was leaving the group. Jordan jumped on the opportunity and was quickly trained, and the rest is history. I encouraged a collaboration with Vanderbilt professor of physics Richard Haglund who specializes in optical physics. Richard was enthusiastic about bringing the spatial resolution of the STEM to study luminescence from nanostructures decorated with plasmonic nanoparticles. In a STEM, in addition to CL, one can also get electron-energy-loss spectra (EELS), which probe electronic excitations. Each of the two methods had been used separately to study surface plasmons, but Jordan's vision was to get spatially resolved CL and EELS on the same sample to exploit the complementary nature of the two techniques: EELS probes beam-induced electronic excitations, while CL probes their radiative decay, which allows one to directly obtain a spatially and spectrally resolved picture of the plasmonic characteristics of nanostructures in three dimensions. He carried out a comprehensive study of surface plasmons of arbitrary morphology of Ag nanoparticles on ZnO/NgO coreshell nanowires and demonstrated his vision.

The connection to Haglund's group led to another important connection, namely, with Ben Lawrie, a former student of Haglund who was a staff member at ORNL. The three-way collaboration led to a project on the generation and detection of orbital angular momentum (OAM) modes in nanostructures. OAM of light is the component of angular momentum of a light beam that is dependent on the field spatial distribution, not the polarization, e.g., a helical wave front. Here again, in STEM-CL, the converged, high-energy electron probe excites optical phenomena across the entire spectrum and provides the highest spatial resolution available for optical spectroscopy, allowing for the study of OAM with simultaneous spatial and spectral resolution. Jordan used STEM-CL to study plasmonic vortices generated by so-called Archimedean spiral channels in silver films. The spatial and spectral resolution of STEM-CL allowed the mapping of the plasmonic vortex with precision and the tracking of its dispersion through the visible regime, demonstrating that the vortex is resolved over a wide spectral range ($\sim$560–660 nm). More importantly, the STEM-CL experiment provided a spatially resolved observation of the vortex phase, the first time that such an observation is made without employing time-resolved measurements.

In addition to his main themes mentioned above, Jordan was always ready to try something else with his magical equipment. At Vanderbilt, I have a collaborative program with engineers studying reliability of electronic devices. Jordan participated in some of this work and provided amazing spatially resolved chemical maps of electronic devices that were very useful in analyzing electrical data that monitor alteration of device properties. Additionally, he was instrumental in providing microstructural and compositional characterization of nanoparticles that combine iron oxide and gold so that both have magnetic and plasmonic properties. This was a collaboration with a group from Barcelona, initiated through a visitor at the ORNL group.

Overall, Jordan had the unusual experience of being mentored by a theorist and several experimentalists in his pursuit of a doctoral thesis. His thesis ended up being mostly experiments for which he was mentored by Steve Pennycook and, after Steve left ORNL, by Matt Chisholm for microscopy and by Richard Haglund and Ben Lawrie on optical and plasmonic physics. He learned how to navigate in a complex environment with a wealth of resources and took advantage of all that was there. His thesis reflects that journey of search and discovery.

Nashville, TN, USA
September 2017

Prof. Sokrates T. Pantelides

Acknowledgments

It will be impossible for me to properly give thanks for all the assistance, encouragement, guidance, mentorship, and inspiration that I received on my path to a Ph.D. That being said, I am going to try.

First, I have to thank Professor Sokrates T. Pantelides, my thesis advisor. The last 5 years has been such an incredible experience, and I am truly grateful for everything he's done for me. Working with him, I have had all the independence I wanted, all the mentorship I needed, and all the opportunities I could've asked for. I am also indebted to all my other committee members, Professor Richard Haglund, Professor Gerd Duscher, Professor Stephen Pennycook, and Professor Kalman Varga for the guidance and direction I received during my time as a student at Vanderbilt.

I also must thank all of the people in the Pantelides group who helped teach me DFT, Yevgeniy Puzyrev, Bin Wang, and especially Xiao Shen. I must single out Rohan Mishra for additional thanks, because he helped me on so many levels, not just on how to perform DFT calculations but on how to be a better scientist in general. I also thank Gerd Duscher and his student Ondrej Dyck with whom I started my first research project, along with Ramki Kalyanaraman, Abhinav Malasi, Phillip Rack, and Yueying Wu.

One of the most important opportunities of my life has been the opportunity to work in Oak Ridge National Laboratory with the STEM Group, and next, I'd like to thank all the people who helped me learn electron microscopy. I thank all the postdocs who both trained me in the lab and were there for fascinating discussions (both scientific and not) Ritesh Sachan, Anas Mouti, Manuel Roldan, Chen Li, and Jon Poplawsky. I thank Bill Sides and Andy Lupini for keeping the microscopes up and running and for teaching me how to take care of a microscope myself (and an extra thanks to Andy for all his help in our collaborations together). I'd also like to thank my current friend and future boss Juan Carlos Idrobo who was also a tremendous mentor during my time here at Oak Ridge. And finally, I thank the two group leaders, Steve Pennycook and Matt Chisholm, who have not just hosted me here and given me an opportunity to work alongside and learn from the best but helped me directly as mentors and collaborators.

The collaborations I have been a part of are also something I am tremendously grateful for. I thank Anna Roig in Barcelona and also her student Siming Yu for the fantastic collaboration with the nanoparticle decorated nanotriangles. I thank the Radiation Effects and Reliability Group in the Electrical Engineering Department and Professors Ronald Schrimpf, Daniel Fleetwood, and Robert Reed, and especially Guoxing Duan, for getting me my first papers in grad school. Also I thank Jérôme Mitard at IMEC for providing me with a chance to actually work hands on with cutting-edge nanotechnology. I must thank Richard Haglund a second time, because through his group at Vanderbilt, I have had several highly productive collaborations, with Claire Marvinney and Daniel Mayo on the ZnO nanowires and with Rod Davidson on a whole number of projects (and who has also been a great friend and my primary source of commiseration during graduate school). Also at ORNL, not just in the STEM Group, but I must thank Benjamin Lawrie in QIS who has recently become a terrific collaborator, and at CNMS, Scott Retterer, Bernadeta Srijanto, and Dale Hensley who have been instrumental in teaching me the nanofabrication process in the clean room. I also have to thank Sang-Yeong Park at New Mexico State University for all his tremendous ideas and insights that have led to some very exciting results.

Finally, I have to thank my family, my sister, Kira, for being my best friend, my mom, Linda, for teaching me how to know myself and grow as a person, my dad, Gary, for being my inspiration for pursuing the study of science at the highest level; and of course my wife, Helen, for keeping me sane and happy, and being there for me every step of the way.

Thank you all, I couldn't have done it without you.

Jordan A. Hachtel

Contents

1 **Introduction** 1
   1.1 Nanoscale Complexity in Modern Nanotechnology 1
   1.2 The Interaction of Light and Matter 4
         1.2.1 Optical Properties in Semiconductors 5
   1.3 Plasmonics: Controlling Light at the Nanoscale 6
         1.3.1 Plasmon Resonances in Semiconductors and Metals 6
         1.3.2 Propagating Plasmon Polaritons at Metal/Dielectric Interfaces 8
         1.3.3 Localized Surface Plasmon Resonances 10
         1.3.4 Surface Plasmons in Complex Nanostructures 11
   References 13

2 **Tools and Techniques** 17
   2.1 Density Functional Theory: Quantum Mechanics for Complex Systems 17
         2.1.1 Calculating Optical Properties with Density Functional Theory 19
   2.2 Finite-Difference Time-Domain: Electrodynamics for Nanostructures 21
   2.3 Scanning Transmission Electron Microscopy: Ultrahigh Resolution Analysis 22
         2.3.1 Correcting Aberrations in an Electron Probe 24
         2.3.2 Bright Field and Dark Field in the STEM 25
   2.4 Electron-Beam Spectroscopies for Nanoscale Optical Properties.... 27
         2.4.1 Electron Beam Interactions with Materials 28
         2.4.2 Electron Energy Loss Spectroscopy 29
         2.4.3 Cathodoluminescence 31
   References 34

3 Extracting Interface Absorption Effects from First-Principles 37
- 3.1 Atomistic Interface Effects 37
  - 3.1.1 Extracting Atomistic Interface Absorption Effects 38
  - 3.1.2 $\Delta\alpha_i$ the Interface Absorbance Difference 40
  - 3.1.3 Accuracy of the Generalized Gradient Approximation 41
  - 3.1.4 Absorption and Reflection at the Atomic Scale 42
- 3.2 Converging the Interface 44
- 3.3 Inverted Design Through Interface Concentration 44
  - 3.3.1 Combining Distinct Interfaces 44
- 3.4 Quantitative Applications 46
  - 3.4.1 Interface Absorption vs. Bulk Absorption 46
  - 3.4.2 Wavelength Selectivity and Absorption Enhancement 47
- References 49

4 Advanced Electron Microscopy for Complex Nanotechnology 53
- 4.1 Ge-Based FET Devices 53
  - 4.1.1 Negative-Bias Temperature Instability in Flat Si-Capped pMOSFETs 54
  - 4.1.2 Structural and Compositional Study of Ge pMOS FinFETs 57
- 4.2 Magnetic and Plasmonic Nanocomposites 61
  - 4.2.1 Composition of Nanocomposite Components 61
  - 4.2.2 Bonding of SPIONs to the Au Nanostructures 65
  - 4.2.3 The Optical Response of the Nanocomposites 69
- References 73

5 Probing Plasmons in Three Dimensions 75
- 5.1 Plasmons in Three-Dimensional Structures 75
- 5.2 Complementary Spectroscopies in the Electron Microscope 76
  - 5.2.1 Surface Plasmons Observed in Both EELS and CL 78
  - 5.2.2 Constant Background Subtraction in EELS Spectrum Imaging 79
  - 5.2.3 Surface Plasmons Observed Only in EELS 82
  - 5.2.4 Surface Plasmons Observed Only in CL 83
- 5.3 Validation of Experimental Results 85
  - 5.3.1 Approximating Nanoparticle Geometries 85
  - 5.3.2 Finite-Difference Time-Domain Confirmation of Experimental Analysis 87
- References 89

6 The Plasmonic Response of Archimedean Spirals 91
- 6.1 Combining Photonics and Electron Microscopy for Plasmonic Analyses 92
  - 6.1.1 EELS Analysis of Lithographically Prepared Nanostructures 93
  - 6.1.2 Enhancing STEM with Photonics 94

6.2	Orbital Angular Momentum in Plasmonic Spiral Holes...........	97
	6.2.1 Visualizing Orbital Angular Momentum with Cathodoluminescence...	99
References...		102

7 Future Directions and Conclusion 105
7.1 Advanced Experiments for Nanoscale Optical Analyses 105
7.2 Outlook and Conclusion 107
References... 107

Appendix A   Overview of Electron Microscopes 109
  A.1 Nion UltraSTEM 200 109
  A.2 VG-HB601... 110
  A.3 Zeiss Libra200-MC... 110
  Reference 111

Appendix B   Fit Parameters EELS and CL Data in Chap. 5 113

Appendix C   Sample Preparation for STEM Analysis 115
  C.1 Solid-State Device Cross-Sections with Dual Beam FIB/SEM 115
  C.2 Direct Sample Preparation of Nanospiral Arrays with EBL......... 117

Abbreviations

ADF	Annular dark field
BF	Bright field
BZ	Brillouin zone
CAD	Computer aided design
CBED	Convergent beam electron diffraction
CL	Cathodoluminescence
DFT	Density functional theory
EBL	Electron-beam lithography
EELS	Electron energy loss spectroscopy
FDTD	Finite-difference time-domain
FFT	Fast Fourier transform
FIB	Focused ion beam
FWHM	Full-width half-maximum
GGA	Generalized gradient approximation
HAADF	High angle annular dark field
iFFT	Inverse fast Fourier transform
LA	Long axis
LDA	Local density approximation
LDOS	Local density of states
LR	Log ratio
LSPR	Localized surface plasmon resonance
MLLS	Multiple linear least squares
NBTI	Negative bias temperature instability
NH	Nanohexagon
NIR	Near-infrared
NT	Nanotriangle
OAM	Orbital angular momentum
OOP	Out of plane
PMT	Photomultiplier tube
PVP	Polyvinylpyrrolidone
RIE	Reactive ion etching

ROI	Region of interest
SA	Short axis
SEM	Scanning electron microscopy
SHG	Second harmonic generation
SI	Spectrum image
SPION	Superparamagnetic iron oxide nanoparticle
SPP	Surface plasmon polariton
STEM	Scanning transmission electron microscopy
TEM	Transmission electron microscopy
TR	Transition radiation
VASP	Vienna ab initio software package
XRPA	X-ray photoabsorption
ZLP	Zero loss peak

Parts of this thesis have been published in the following journal articles:

1. Hachtel JA, Davidson II RB, Lupini AR, Lawrie BJ, Haglund RF, Pantelides ST. Near-field polarization selectivity of complex modes in plasmonic Archimedean nanospirals through cathodoluminescence imaging. In Preparation.
2. Hachtel JA, Cho SY, Davidson II RB, Chisholm MF, Haglund RF, Idrobo JC, Pantelides ST, Lawrie BJ (2017) Spatially and spectrally resolved orbital angular momentum interactions in plasmonic vortex generators. Preprint, arXiv:1705.10640.
3. Zhang EX, Fleewood DM, Hachtel JA, Liang C, Reed RA, Alles ML, Schrimpf RD, Linten D, Mitard J, Chisholm MF, Pantelides ST (2017) Total ionizing dose effects on Ge pMOS FinFETs on bulk Si. IEEE Transactions on Nuclear Science 64:226.
4. Hachtel JA, Yu S, Lupini AR, Pantelides ST, Gich M, Laromaine A, Roig A (2016) Gold nanotriangles decorated with superparamagnetic iron oxide nanoparticles: a compositional and microstructural study. Faraday Discussions 191:215.
5. Hachtel JA, Marvinney C, Mouti A, Mayo DC, Mu R, Pennycook SJ, Lupini AR, Chisholm MF, Haglund RF, Pantelides ST (2016) Probing plasmons in three dimensions by combining complementary spectroscopies in an scanning transmission electron microscope. Nanotechnology 27:155202.
6. Yu S, Hachtel JA, Chisholm MF, Pantelides ST, Laromaine A, Roig A (2015) Magnetic gold nanotriangles by microwave-assisted polyol synthesis. Nanoscale 7:14039.
7. Duan GX, Hachtel J, Shen X, Zhang EX, Zhang CX, Tuttle BR, Fleetwood DM, Schrimpf RD, Reed RA, Franco J, Linten D, Mitard J, Witters L, Collaert N, Chisholm MF, Pantelides ST (2015) Activation energies for oxide- and interface-trap charge generation due to negative-bias temperature stress of Si-capped SiGe-pMOSFETs. IEEE Transactions on Device and Materials Reliability 15:352.
8. Hachtel JA, Sachan R, Mishra R, Pantelides ST (2015) Quantitative first-principles theory of interface absorption effects in multilayer heterostructures. Applied Physics Letters 107:091908.

Chapter 1
Introduction

1.1 Nanoscale Complexity in Modern Nanotechnology

The drive for miniaturization has pushed nanotechnology to the forefront of the materials science community. Perhaps the most famous example has been Moore's law, the prediction by G.E. Moore that the number of transistors in an integrated circuit would double every 2 years [1]. However, the desire for devices with real-world applications and increasingly small dimensions extends far past transistors, as miniaturization has become a key aspect across many subfields of science [2–6].

As device dimensions push into the nanoscale, one of the main focuses of the research community has been on the interactions of light and matter. Optical nanostructures are of significant interest across a wide range of technological subfields such as photovoltaics [7–10], biomedicine [11, 12], catalysis [13–15], sensing and detection [16–18], laser optics [19–21], and optoelectronics [22–25].

As devices have pushed deeper and deeper into the nanoscale, they have encountered new regimes where complex physical phenomena that were dormant at the micro and macroscales rear their heads [12, 26–28]. The result has been an increased research effort into nanoscale optical effects that has resulted in parallel endeavors in the fields of nanoscale fabrication, and tremendous advances have been made in terms of colloidal synthesis [29–33], lithography [34–37], thin film deposition [38–40], self-assembly [41–44], and focused ion beam (FIB) techniques [45–49]. As the control over materials in fabrication has increased, so has the precision required for device applications.

Figure 1.1 shows several examples of nanostructures with a strong dependence on geometry and morphology. In each of these examples, novel applications are generated from precision control of the nanoscale features of complex nanostructures. They demonstrate that to truly understand the full complexity of modern

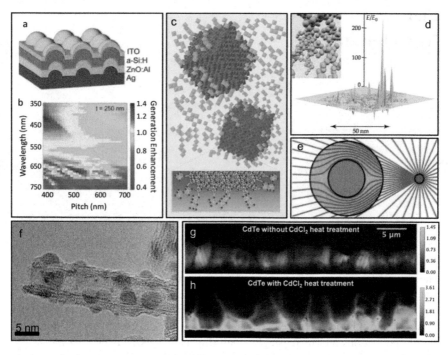

Fig. 1.1 Complex nanostructures in modern nanotechnology. (**a**) Heterostructure solar cells with plasmonic nanopatterning on the back contact. (Figures reprinted with permission from [50]. Copyright 2011 American Chemical Society.) (**b**) The pitch of the nanopatterning in (**a**) significantly affects both the intensity and wavelength of peak charge-carrier generation in the sample. (**c**) ITO nanocrystals embedded in amorphous NbO_x. The bonding between the NbO_x and the ITO changes the optical response under bias, allowing for the formation of smart windows. (Reprinted by permission from Macmillan Publishers Ltd: Nature [51], copyright 2013.) (**d**) Aggregates of Ag nanoparticles dispersed randomly show a high environmental dependence on the plasmonic electric field enhancement. (Figure reproduced from [52] with the permission of the American Institute of Physics.) (**e**) By creating metamaterials with negative indices of refraction, light can be bent around an object allowing light to travel past it, as if it were not there, to allow for optical cloaking. (From [53]. Reprinted with permission from AAAS.) (**f**) TiO_2 nanotubes decorated with PbS quantum dots. The functionalization of the nanotube with quantum dots significantly enhances photocatalytic activity and stability while increasing the usable portion of the visible spectrum. (Reprinted with permission from [54]. Copyright 2009 American Chemical Society.) (**g**) CdTe solar cells lose efficiency at the grain boundaries, but by altering the atomic scale composition of the grain boundary with chemical treatments, (Reprinted figure from [55]: C. Li et al., Phys. Rev. Lett. **112**, 156103 (2014). Copyright 2014 American Physical Society.) (**h**) the grain boundaries can generate even stronger photocurrent than the bulk CdTe

nanotechnology, we must understand it *at the nanoscale*. This is the focus of this dissertation: to study nanoscale physical, and specifically optical, phenomena that occur in complex nanostructures, and to directly study them at the nanoscale through both experimental and theoretical techniques.

Many of these optical phenomena arise from the fundamental interactions between photons and the valence electrons in a solid at the quantum mechanical level. At the atomistic level, the optical response is dictated by the band structure and the dielectric function, and in the remainder of this first introductory chapter I will introduce the origin and formula for the dielectric function, as well as many of the different optical phenomena that can affect the utility of complex nanostructures. The quantum mechanical picture explains many important optical phenomena; however the majority of this dissertation will focus on a more classical optical phenomenon: the plasmon. In the last section of this chapter, I introduce what a plasmon is and show several different types of plasmonic resonances that have a wide range of applications in modern technology.

Chapter 2 introduces the tools and techniques required to study the optical phenomena from Chap. 1 *at the nanoscale*. From the theoretical side, I introduce different types of calculations that can access both the quantum mechanical and classical optical response of complex nanostructures. The bulk of the work in this dissertation is experimental, and utilizes electron microscopy and related spectroscopies, which can directly detect and measure both the geometry and the near-field response of complex nanostructures.

After the concepts and techniques are introduced, the analyses of complex optical nanostructures can begin. In Chap. 3, I approach the problem of understanding absorption in multilayer heterostructures from a purely theoretical standpoint. In order to determine the genuine response of a large scale heterostructure with multiple interfaces, one must first understand the effects of the unique interface bonds on the absorption spectrum of the structure. In this chapter I provide a means of quantitatively determining the interface absorption using quantum mechanical calculations and then show how such calculations can be applied toward rationally designing heterostructures with selectable absorption profiles.

The experimental research in this dissertation focuses on direct observation of nanoscale effects using electron microscopy. Chapter 4 shows a series of different analytical techniques within a microscope that can be used to characterize different complex nanostructures. The results here demonstrate the wide breadth and versatility of electron microscopy as a tool for nanoscale analyses.

In the final two chapters, I will focus on plasmonics in complex nanostructures. Utilizing electron microscopy to study plasmonics is a powerful combination because the associated spectroscopies allow direct access to the nanoscale near-field response of plasmonic structures. Chapter 5 demonstrates how complimentary spectroscopies in the electron microscope can be used to collect three-dimensional data from plasmonic samples without the use of tomography, simulation, or computerized reconstructions. Then in Chap. 6, nanoscale spiral structures with a complicated plasmonic response are studied with the electron microscope. The complexity of the plasmonic response as well as novel physical phenomena is unveiled by combining high spatial resolution electron microscopy with photonic techniques and computer simulation.

The final chapter is the future directions and conclusion of the dissertation. I show here preliminary data from a new generation of experiments that I will perform in the next step of my research career. And lastly I provide a perspective on this body of work as a whole and where it fits into the world of nanotechnology.

1.2 The Interaction of Light and Matter

To utilize light toward advanced applications, a quantitative way of understanding the manner in which light and matter interact with one another is needed. Since light is an electromagnetic wave, its behavior in a material is defined by Maxwell's equations

$$\nabla \cdot \epsilon_0 \epsilon_r \mathbf{E} = \rho \tag{1.1}$$

$$\nabla \times \mathbf{E} = -\frac{\partial \mathbf{B}}{\partial t} \tag{1.2}$$

$$\nabla \cdot \mathbf{B} = 0 \tag{1.3}$$

$$\nabla \times \frac{\mathbf{B}}{\mu_0 \mu_r} = \mathbf{j} + \epsilon \frac{\partial \mathbf{E}}{\partial t} \tag{1.4}$$

Where $\mathbf{E}$ and $\mathbf{B}$ are the electric and magnetic fields, ρ and $\mathbf{j}$ are the current and charge densities, and ϵ_0 and μ_0 are the permittivity and the permeability in free space while ϵ_r and μ_r are the relative permittivities and permeabilities of the material. It is important to note that the permittivity and permeability of a material are not independent variables, as they are connected by the relation,

$$\epsilon_r \mu_r = n^2 \tag{1.5}$$

where n is the refractive index of the material, defined as the ratio between the speed of light in a vacuum and the speed of light in the material. This dissertation primarily deals with nonmagnetic materials, for which $\mu_r = 1$. Thus, the relative permittivity determines the index of refraction and is the primary parameter that controls the optical response of the material.

However, the relative permittivity is a function of both the momentum, $\mathbf{q}$, and the frequency, ω, of the incident light. The quantity is better described as the dielectric function, and a formula to determine the dielectric function in connection to band theory was developed in a series of papers by Ehrenreich, Cohen, Adler, and Wiser in the 1950s and 1960s [56–58]. The expression gives the full three-dimensional dielectric response of a crystal in terms of $\mathbf{q}$ and ω. It is a matrix in the crystal's reciprocal lattice vectors, $\mathbf{G}$:

$$\epsilon_{G,G'}(\mathbf{q},\omega) = \delta_{G,G'} - \frac{8\pi e^2}{\Omega |\mathbf{G}+\mathbf{q}||\mathbf{G}'+\mathbf{q}|} \sum_{c,v,\mathbf{k}} \frac{\langle \psi_{c\mathbf{k}+\mathbf{q}}| e^{i(\mathbf{q}+\mathbf{G})\cdot\mathbf{r}} |\psi_{v\mathbf{k}}\rangle \langle \psi_{v\mathbf{k}}| e^{-i(\mathbf{q}+\mathbf{G}')\cdot\mathbf{r}'} |\psi_{c\mathbf{k}+\mathbf{q}}\rangle}{E_{c\mathbf{k}+\mathbf{q}} - E_{v\mathbf{k}} - \omega} \tag{1.6}$$

1.2 The Interaction of Light and Matter

Here, c, υ, and $\mathbf{k}$ are all indices referring to different points in the band structure, c and υ refer to the index of bands in the conduction and valence band, while $\mathbf{k}$ refers to a point in the Brillouin zone (BZ), $\psi_{c,\upsilon\mathbf{k}}$ and $E_{c,\upsilon\mathbf{k}}$ are the Bloch functions and associated energy eigenvalues for the system, and Ω is the volume of the primitive unit cell.

Furthermore, since a Bloch function is the product of a plane wave and a cell-periodic function, $|\psi_{\upsilon\mathbf{k}}\rangle = e^{i\mathbf{k}\cdot\mathbf{r}}|u_{\upsilon\mathbf{k}}\rangle$, the numerator of the fraction at the end of Eq. (1.6) can be rewritten as

$$\langle\psi_{c\mathbf{k}+\mathbf{q}}|e^{i\mathbf{q}\cdot\mathbf{r}}|\psi_{\upsilon\mathbf{k}}\rangle\langle\psi_{\upsilon\mathbf{k}}|e^{-i\mathbf{q}\cdot\mathbf{r}'}|\psi_{c\mathbf{k}+\mathbf{q}}\rangle = |\langle u_{c\mathbf{k}+\mathbf{q}}||u_{\upsilon\mathbf{k}}\rangle|^2 \quad (1.7)$$

From the above examinations, one can infer that the $\mathbf{q}$-dependence derives from the overlap between the cell-periodic Bloch functions of the initial and final states, which imposes selection rules on the transitions. The ω-dependence essentially functions as a Dirac delta function. Where the energy of the incoming photon, $\hbar\omega$, must be equal to the difference between the final and initial states of the electron in order to conserve energy.

As a result, the dependence of the dielectric function on $\mathbf{q}$ and ω can be described qualitatively as a weighted sum of transitions from the valence to the conduction bands.

1.2.1 Optical Properties in Semiconductors

The dielectric function is important to understand the optical properties of materials, and especially semiconductors. The expression in Eq. (1.6) determines the optical response via band-to-band (interband) transitions. The lowest available interband transition determines the bandgap, which is a defining characteristic of the optical response of a semiconductor. Figure 1.2a shows a schematic of basic interband absorption and emission in a semiconductor. Light with energy at least equal to the bandgap can be absorbed and excite an electron from the valence to the conduction band, and the shape of the absorption spectrum is determined by the number of allowed transitions at each energy.

Once excited the electron begins to relax back to ground state by releasing energy in the form of photons or phonons. Electrons trickle down to the conduction band minimum by emitting phonons, because photons can only be emitted by $q = 0$ transitions. In a direct-gap material, photons are emitted across the bandgap, but for materials with an indirect bandgap, relaxation to the valence band occurs either by a phonon-assisted process or an Auger process.

Interband excitations are not the whole story, even in the case of a perfectly crystalline materials. In semiconductors, a class of optical excitations called excitons can be formed, which are frequently represented as a conduction band electron and its vacant electron-hole in the valence band coupled to one another through Coulomb attraction. The pair is considered a quasiparticle and essentially behaves as a

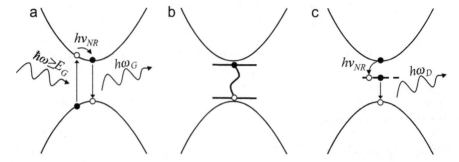

Fig. 1.2 Optical excitations in semiconductors. (**a**) A semiconductor allowing absorption from the conduction to the valence band for energies above the gap and emitting light with energy equal to the bandgap energy. (**b**) The effect of defects and dopants is to add new electronic states in the bandgap, altering both the absorption and emission properties. (**c**) Excitons, bound electron-hole pairs, function as a quasiparticle excitation in semiconductors, significantly changing the behavior of excited carriers

hydrogenic atom, with a significantly lower effective mass (schematic in Fig. 1.2b). Excitons can have a strong effect on the absorption and emission spectrum of a material and can also influence transport properties in materials, making them highly important in the world of optical devices [59, 60].

Additionally, disruptions to the crystalline structure can significantly change the optical response of a semiconductor. Defects or impurities in the crystal essentially create localized energy levels in the band structure. These "defect levels" open new transitions to the electrons which provide alternative paths for electron-hole recombination (Fig. 1.2c). Defects and impurities can also function as non-radiative traps and recombination centers that negatively impact devices, but in some cases the defect levels can form stable highly luminescent states used to control the optical response of the material [61–63].

1.3 Plasmonics: Controlling Light at the Nanoscale

In additions to those discussed in the last section, there are many more types of fundamental optical excitations. One in particular bears significant relevance to this dissertation and warrants a deeper introduction: the plasmon. There are many different forms of plasmons, and here I will discuss them briefly.

1.3.1 Plasmon Resonances in Semiconductors and Metals

The plasmon is a many-body excitation of the conduction band electrons. The origin of the excitation can be seen in the energy-loss function, which describes

1.3 Plasmonics: Controlling Light at the Nanoscale

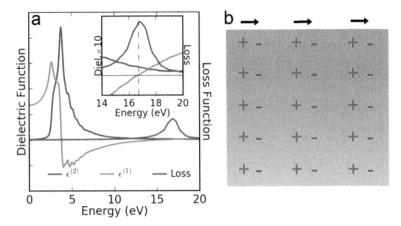

Fig. 1.3 Bulk plasmon resonances in semiconductors and metals. Bulk plasmons occur when the real part of the dielectric function is zero, and the imaginary part is small. (**a**) The dielectric and loss functions of Si, showing the origin of a bulk plasmon resonance in the loss function at 16.7 eV. (**b**) The physical interpretation is that the electrons (red minuses) in the material are all uniformly displaced from the lattice ions (blue plusses). The result is a uniform electric field (black arrows) between the lattice and the electron density that generates a restoring force, and creates an oscillation of the charge density around the lattice ions

the response of a solid to an incoming fast electron

$$\text{Loss} = \text{Im}\left[-\frac{1}{\epsilon}\right] = \frac{\epsilon_2}{\epsilon_1^2 + \epsilon_2^2}, \quad \epsilon = \epsilon_1 + i\epsilon_2 \quad (1.8)$$

where ϵ_1 and ϵ_2 are the real and imaginary parts of the dielectric functions. When the real part of the dielectric function is equal to zero, and the imaginary part is very small, a resonance occurs. This is a plasmonic excitation.

Figure 1.3a shows the real and imaginary components of the dielectric function of Si. Most of the features of the dielectric spectrum occur in the proximity of the bandgap (~2–6 eV), but by examining the loss function, one can see that the plasmon resonance occurs at 16.7 eV and represents a very different kind of excitation. To better understand the physical nature of plasmonic resonances, we turn to Drude theory.

The Drude model assumes that conduction electrons in a metal move freely with respect to the ions, meaning the electrons can be described as a free-electron gas [64]. As a result, if an external field (perhaps due to an impinging electron or photon) was to displace a large portion of conduction electrons, an electric field between the static ions and the displaced electrons would generate a restoring force, as is shown in the schematic in Fig. 1.3b, and the electron oscillations can be treated as a harmonic oscillator

$$m_0 \frac{d^2 x(t)}{dt^2} + m_0 \gamma \frac{dx(t)}{dt} = -e\left[E(t) = E_0 e^{-i\omega t}\right] \rightarrow x(t) = \frac{eE(t)}{m_0} \frac{1}{\omega^2 + i\gamma\omega} \quad (1.9)$$

where m_0 is the mass of the electron, E_0 is the amplitude of the restoring electric field, and γ is a damping term. The Drude model can be used to create a new expression for the dielectric function by considering an alternate expression for Eq. (1.1) in terms of the polarization, $P(t) = -Nex(t)$

$$\epsilon_0\epsilon_r E(t) = \epsilon_0 E(t) + P(t) \rightarrow \epsilon_r(\omega) = 1 - \frac{\omega_p^2}{\omega^2 + i\gamma\omega} \quad (1.10)$$

where

$$\omega_p = \sqrt{\frac{Ne^2}{m_0\epsilon_0}} \quad (1.11)$$

is called the plasma frequency, N is the number of free carriers, and m_0 is the weight of the electron.

The origin of bulk plasmon resonances can be understood by replacing ϵ_r in Eq. (1.1) with the expression in Eq. (1.10) and considering a neutral metal (no net charge), Eq. (1.1) becomes

$$\epsilon_0\left(1 - \frac{\omega_P^2}{\omega^2 + i\gamma\omega}\right)\nabla \cdot \vec{E}(x) = 0. \quad (1.12)$$

For most frequencies, Eq. (1.12) requires that $\nabla \cdot \vec{E}(x) = 0$. However, for lightly damped materials where $\gamma \approx 0$, $\omega = \omega_p$ satisfies the equation, which allows for the creation of longitudinal oscillating modes in the electron gas of the solid.

Equation (1.8) and Eqs. (1.10) and (1.12) present two different physical pictures for the origin of the bulk plasmon, and while Drude theory is specifically designed for treating metals, the bulk plasmon mode is present in semiconductive and insulator materials as well. As will be discussed in the next section, the bulk plasmon is not an optical phenomenon, since it is a longitudinal mode and cannot in-couple or out-couple to a transverse wave like a photon. However, the bulk plasmon can be excited by an electron, and is important to understand when performing analyses with an electron microscope.

1.3.2 Propagating Plasmon Polaritons at Metal/Dielectric Interfaces

Plasmons, while not optical themselves, can interact with light, and when an electromagnetic wave couples to a plasmon, it forms a polariton and propagates through the medium. Polaritons must satisfy the wave equation

$$\nabla^2\mathbf{E}(\mathbf{r},t) - \epsilon_r\frac{1}{c^2}\frac{\partial^2\mathbf{E}(\mathbf{r},t)}{\partial t^2} = 0 \quad (1.13)$$

1.3 Plasmonics: Controlling Light at the Nanoscale

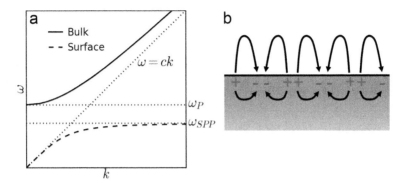

Fig. 1.4 Polariton propagation and dispersion. (**a**) The plasmon polariton dispersion for bulk plasmons (solid) and surface plasmon polaritons (SPPs—dashed). (**b**) Schematic of field lines and charge distribution during SPP propagation

If treated as a standard propagating wave, $\mathbf{E}(\mathbf{r}, t) = \mathbf{E}(\mathbf{k}, \omega)e^{i(\mathbf{k}\cdot\mathbf{r}-\omega t)}$, and ϵ_r is taken from Eq. (1.10), the relationship between $\mathbf{k}$ and ω (neglecting the damping term) is required to be

$$\omega^2 = \omega_p^2 + c^2 k^2. \tag{1.14}$$

This relation describes the bulk plasmon polariton dispersion curve and is plotted in Fig. 1.4a. An important aspect about this dispersion curve is that if ω is less than ω_P, the required wavevector becomes imaginary, and plasmon excitation is not possible. As a result, bulk plasmons are usually not an important part of optical experiments, as the required frequencies are far out of the visible regimes.

However, the interface between two materials which have opposite signs on the real part of the dielectric function, specifically a metal (negative ϵ_1) and a dielectric (positive ϵ_1), can support surface plasmons which often have lower frequencies to allow for strong optical interactions. The surface plasmons are much like bulk plasmons, in that they consist of collective oscillations of electron charge, but differ in the sense that the surface plasmon resonances are highly confined to the interface with the dielectric mismatch.

Plasmon polaritons can also exist at such interfaces in the form of surface plasmon polaritons (SPP), which have a broad range of frequencies. A schematic of the charge and field response of a propagating SPP wave is shown in Fig. 1.4b.

Just like the surface plasmon, the SPP is tightly confined to the interface. As a result, the form for the propagating wave gets another term, $\mathbf{E}_{SPP}(\mathbf{r}, t) = \mathbf{E}_{SPP}e^{-\kappa|z|}(\mathbf{k}, \omega)e^{i(\mathbf{k}\cdot\mathbf{r}-\omega t)}$, where κ is some damping constant depending on the material (different for the metal and dielectric sides) and $|z|$ is the absolute distance away from the interface.

By applying the boundary conditions at the interface and solving Maxwell's equations, the dispersion of the SPP can be shown to be

$$k = \frac{\omega}{c}\sqrt{\frac{\epsilon_d(\omega^2 - \omega_P^2)}{(1+\epsilon_d)\omega^2 - \omega_P^2}} \tag{1.15}$$

where ϵ_d is dielectric function of the dielectric material. This dispersion is also plotted in Fig. 1.4a and can be seen to capture the range of frequencies going down to zero. It is important to note that the $\omega - k$ relationship for all photons is defined by the light line, $\omega = ck$. The dispersion curves of the SPP (and the bulk plasmon polariton) both asymptotically approach the line but never reach it, meaning light cannot be used to directly excite either. However, by patterning metal surfaces with periodic gratings, transverse light can be coupled in and out of the longitudinal SPP modes very efficiently [65, 66]. The coupled system is highly versatile, and allows for the control of light with confinement of the electric field to the interface on the scale of nanometers, and propagation distances for waveguiding on the order of hundreds of microns [50, 67, 68].

As an alternative to patterning the sample, fast electrons can be used to directly excite the SPPs in materials [69]. When a fast-moving electron transitions between a dielectric environment and a metallic environment, the induced electric field in the two media requires additional fields to be created at the interface for the boundary conditions to be satisfied. For an electron traveling in the $\hat{z}$ direction and passing from material α to material β, one would require the boundary conditions $\epsilon_\alpha(E_{z\alpha} + E'_{z\alpha}) = \epsilon_\beta(E_{z\beta} + E'_{z\beta})$, $E_{r\alpha} + E'_{r\alpha} = E_{r\beta} + E'_{r\beta}$, and $k'_r E'_r + k'_z E'_z = 0$, where E is the induced field of the electron, E' is the additional field generated by the fast electron to satisfy the boundary conditions, and k' is the wavevector of the generated electromagnetic wave. The additional electromagnetic wave can couple with the surface plasmon to form the SPP [70]. This allows for electron-beam techniques to both excite and probe SPP effects in complex nanostructures across a broad range of wavelengths [71, 72].

1.3.3 Localized Surface Plasmon Resonances

Surface plasmon polaritons allow for excellent control of light on quasi-infinite planar surfaces, but for finite structures, with hard boundaries, a different treatment is needed. In 1908, Mie developed a solution to Maxwell's equations for metallic nanospheres that predicted a plasmonic resonance in the visible regime [73]. Under his formulation, the extinction cross section (fraction of light either absorbed or scattered) of the nanosphere is

$$\sigma = \frac{24\pi^2 R^3 \varepsilon_m^{3/2}}{\lambda} \frac{\varepsilon_2}{(\varepsilon_1 + 2\varepsilon_m)^2 + \varepsilon_2^2} \tag{1.16}$$

where ε_1 and ε_2 are the real and imaginary parts of the dielectric function of the metallic nanoparticle, and ε_m is the dielectric constant of the surrounding medium [74]. In Eq. (1.16) it can be seen that the resonance occurs when a specific dielectric match between the metal and the surrounding medium is met, $\varepsilon_1 = -2\varepsilon_m$. Figure 1.5a shows extinction cross sections for 100- nm diameter spheres made of

1.3 Plasmonics: Controlling Light at the Nanoscale

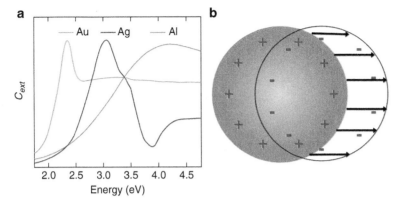

Fig. 1.5 Localized surface plasmon resonances. When Maxwell's equations are solved for metallic spheres, a resonance in the extinction cross section appears when $\epsilon_1 = -2\epsilon_m$, where ϵ_1 is the real part of the metal's dielectric function, and the ϵ_m is the dielectric function of the surrounding medium. This resonance is an LSPR. (**a**) Shows the LSPRs for 100- nm spheres of Au, Ag, and Al in air and how the different materials can significantly alter the plasmonic response. (**b**) Schematic for LSPR oscillations, unlike bulk plasmons or SPPs, where the plasmon propagates through the structure, for LSPRs the entire surface charge of the structure oscillates simultaneously

Au, Ag, and Al in air calculated in Mie theory. The three materials have plasmon resonances ranging from the visible to the IR, demonstrating that the dielectric properties of the material can significantly affect the plasmonic response. (Mie calculations performed using the code available at https://github.com/gevero/py_gmm [75]).

These plasmon resonances are called localized surface plasmon resonances (LSPR). They are localized, because unlike polaritons, they do not propagate, and are confined to the specific volume in which they are excited. Figure 1.5b shows the schematic of the LSPR; here the entire surface charge of the nanoparticle or nanostructure oscillates simultaneously at the LSPR frequency. The strength of these modes is in their extremely high field enhancement at the surface of the nanoparticle [76], strong radiative recombination [17], and high degree of tunability [77].

1.3.4 Surface Plasmons in Complex Nanostructures

Surface plasmons possess strong potential for applications in optical nanotechnology, first and foremost due to their tunability. The plasmon resonances in nanostructures can be tuned by geometry, size, material, and dielectric environment. Figure 1.6a from [78] shows the wide range of different plasmonic materials and the different uses and spectral regimes they represent.

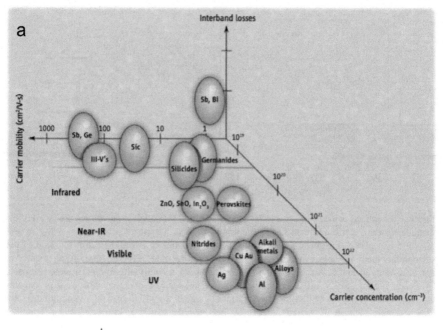

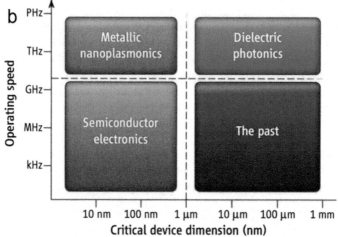

Fig. 1.6 The benefits of plasmonic structures. (**a**) Different materials used for plasmonics shift the utility and spectral range of the plasmon resonances dramatically, opening a wide array of possible applications. (From [78]. Reprinted with permission from AAAS.) (**b**) Additionally, plasmonics combine a unique blend of miniaturizability, and ability to process information at high speed, that separates plasmonics from other phenomena (From [79]. Reprinted with permission from AAAS.)

Furthermore, plasmons can function as antennas and receive and transmit optical signals, opening up the possibility for plasmonic circuits. Figure 1.6b from [79] shows the different regimes in terms of device size and speed. Semiconductors suffer from resistive heating that inherently limits device performance, and photonic devices are diffraction limited and cannot be reduced to nanometer length scales. Plasmonic devices can use light instead of electrons to carry the signal and achieve the high throughput of photonics while confining that light tightly to the surface of nanostructures and operating at the size scales of modern semiconductor nanotechnology.

Surface plasmons, as well as the other optical phenomena discussed in this chapter, provide an excellent toolbox for controlling light at the nanoscale. As a result, methods that can access and analyze these phenomena at the nanoscale have become highly important in the field of nanotechnology. The next chapter is an overview of the tools and techniques (theoretical and experimental) used in this dissertation in order to study the optical properties of complex nanostructures *at the nanoscale*.

References

1. Moore, G.E.: Cramming more components onto integrated circuits. Electronics **38**, 114–116 (1965)
2. Peercy, P.S.: The drive to miniaturization. Nature **406**, 1023–1026 (2000)
3. Huang, Y., et al.: Logic gates and computation from assembled nanowire building blocks. Science **294**, 1313–1317 (2001)
4. Jensen, K.F.: Microreaction engineering—is small better? Chem. Eng. Sci. **56**, 293–303 (2001)
5. Skumryev, V., et al.: Beating the superparamagnetic limit with exchange bias. Nature **423**, 850–853 (2003)
6. Dittrich, P.S., Manz, A.: Lab-on-a-chip: microfluidics in drug discovery. Nat. Rev. Drug Discov. **5**, 210–218 (2006)
7. Tian, B., et al.: Coaxial silicon nanowires as solar cells and nanoelectronic power sources. Nature **449**, 885–889 (2007)
8. Atwater, H.A., Polman, A.: Plasmonics for improved photovoltaic devices. Nat. Mater. **9**, 205–213 (2010)
9. Mishchenko, E.G., Shytov, A.V., Silvestrov, P.G.: Guided plasmons in graphene p-n junctions. Phys. Rev. Lett. **104**, 156806 (2010)
10. Bernardi, M., Palummo, M., Grossman, J.C.: Extraordinary sunlight absorption and one nanometer thick photovoltaics using two-dimensional monolayer materials. Nano Lett. **13**, 3664–3670 (2013)
11. West, J.L., Halas, N.J.: Engineered nanomaterials for biophotonics applications: improving sensing, imaging, and therapeutics. Annu. Rev. Biomed. Eng. **5**, 285–292 (2003)
12. De Angelis, F., et al.: Breaking the diffusion limit with super-hydrophobic delivery of molecules to plasmonic nanofocusing SERS structures. Nat. Photon. **5**, 682–687 (2011)
13. Kamat, P.V.: Photophysical, photochemical and photocatalytic aspects of metal nanoparticles. J. Phys. Chem. B **106**, 7729–7744 (2002)
14. Kamat, P.V.: Meeting the clean energy demand: nanostructure architectures for solar energy conversion. J. Phys. Chem. C **111**, 2834–2860 (2007)
15. Tilley, S.D., Cornuz, M., Sivula, K., Grätzel, M.: Light-induced water splitting with hematite: improved nanostructure and iridium oxide catalysis. Angew. Chem. **122**, 6549–6552 (2010)

16. O'Carroll, D.M., Hofmann, C.E., Atwater, H.A.: Conjugated polymer/metal nanowire heterostructure plasmonic antennas. Adv. Mater. **22**, 1223–1227 (2010)
17. Novotny, L., van Hulst, N.: Antennas for light. Nat. Photon. **5**, 83–90 (2011)
18. Knight, M.W., Sobhani, H., Nordlander, P., Halas, N.J.: Photodetection with active optical antennas. Science **332**, 702–704 (2011)
19. Tredicucci, A., et al.: A multiwavelength semiconductor laser. Nature **396**, 350–353 (1998)
20. Pavesi, L., Dal Negro, L., Mazzoleni, C., Franzò, G., Priolo, F.: Optical gain in silicon nanocrystals. Nature **408**, 440–444 (2000). ISSN: 0028-0836
21. Oulton, R.F., et al.: Plasmon lasers at deep subwavelength scale. Nature **461**, 629–632 (2009). ISSN: 0028-0836
22. Arsenault, A., et al.: Towards the synthetic all-optical computer: science fiction or reality? J. Mater. Chem. **14**, 781–794 (2004)
23. Qi, M., et al.: A three-dimensional optical photonic crystal with designed point defects. Nature **429**, 538–542 (2004)
24. Pillai, S., Catchpole, K.R., Trupke, T., Zhang, G., Zhao, J.: Enhanced emission from Si-based light-emitting diodes using surface plasmons. Appl. Phys. Lett. **88** (2006). ISSN: 0003-6951. https://doi.org/10.1063/1.2195695
25. Di Benedetto, F., et al.: Patterning of light-emitting conjugated polymer nanofibres. Nat. Nanotechnol. **3**, 614–619 (2008)
26. Kish, L.B.: End of Moore's law: thermal (noise) death of integration in micro and nano electronics. Phys. Lett. A **305**, 144–149 (2002)
27. Gramotnev, D.K., Bozhevolnyi, S.I.: Plasmonics beyond the diffraction limit. Nat. Photon. **4**, 83–91 (2010). ISSN:1749-4885
28. Esteban, R., Borisov, A.G., Nordlander, P., Aizpurua, J.: Bridging quantum and classical plasmonics with a quantum-corrected model. Nat. Commun. **3**, 825 (2012)
29. Konstantatos, G., Sargent, E.H.: Nanostructured materials for photon detection. Nat. Nanotechnol. **5**, 391–400 (2010). ISSN:1748-3387
30. Mark, A.G., Gibbs, J.G., Lee, T.-C., Fischer, P.: Hybrid nanocolloids with programmed three-dimensional shape and material composition. Nat. Mater. **12**, 802–807 (2013). ISSN: 14761122
31. Milliron, D.J., Buonsanti, R., Llordes, A., Helms, B.A.: Constructing functional mesostructured materials from colloidal nanocrystal building blocks. Acc. Chem. Res. **47**, 236–246 (2014). ISSN: 0001-4842
32. Akselrod, G.M., et al.: Probing the mechanisms of large Purcell enhancement in plasmonic nanoantennas. Nat. Photon. **8**, 835–840 (2014). ISSN: 1749-4885
33. Hachtel, J.A., et al.: Gold nanotriangles decorated with superparamagnetic iron oxide nanoparticles. Faraday Discuss. **191**, 215–227 (2016)
34. Xia, Y., Whitesides, G.M.: Soft lithography. Annu. Rev. Mater. Sci. **28**, 153–184 (1998)
35. Ozbay, E.: Plasmonics: merging photonics and electronics at nanoscale dimensions. Science **311**, 189–193 (2006). ISSN: 0036-8075, 1095-9203
36. Haberfehlner, G., et al.: Correlated 3d nanoscale mapping and simulation of coupled plasmonic nanoparticles. Nano Lett. **15**, 7726–7730 (2015). ISSN: 1530-6984
37. Bosman, M., et al.: Encapsulated annealing: enhancing the plasmon quality factor in lithographically-defined nanostructures. Sci. Rep. **4** (2014). ISSN: 2045-2322. https://doi.org/10.1038/srep05537. http://www.nature.com/articles/srep05537
38. Wang, H., You, T., Shi, W., Li, J., Guo, L.: Au/TiO2/Au as a plasmonic coupling photocatalyst. J. Phys. Chem. C **116**, 6490–6494 (2012). ISSN: 1932-7447
39. Nishinaga, J., Kawaharazuka, A., Onomitsu, K., Horikoshi, Y.: High-absorption-efficiency superlattice solar cells by excitons. Jpn. J. Appl. Phys. **52**, 112302 (2013). ISSN: 1347-4065
40. Zhang, X., et al.: Optical absorption in InP/InGaAs/InP double-heterostructure nanopillar arrays for solar cells. Appl. Phys. Lett. **104**, 061110 (2014). ISSN: 0003-6951, 1077-3118
41. Habibi, Y., Lucia, L.A., Rojas, O.J.: Cellulose nanocrystals: chemistry, selfassembly, and applications. Chem. Rev. **110**, 3479–3500 (2010)
42. Tan, H., Santbergen, R., Smets, A.H.M., Zeman, M.: Plasmonic light trapping in thin-film silicon solar cells with improved self-assembled silver nanoparticles. Nano Lett. **12**, 4070–4076 (2012). ISSN: 1530-6984

43. Kuzyk, A., et al.: DNA-based self-assembly of chiral plasmonic nanostructures with tailored optical response. Nature **483**, 311–314 (2012). ISSN: 00280836
44. Teulle, A., et al.: Multimodal plasmonics in fused colloidal networks. Nat. Mater. **14**, 87–94 (2015). ISSN: 14761122
45. Matsui, S., et al.: Three-dimensional nanostructure fabrication by focused-ion-beam chemical vapor deposition. J. Vac. Sci. Technol. B **18**, 3181–3184 (2000)
46. Kuttge, M., García de Abajo, F.J., Polman, A.: Ultrasmall mode volume plasmonic nanodisk resonators. Nano Lett. **10**, 1537–1541 (2009). ISSN: 1530-6984
47. Kuttge, M., Vesseur, E.J.R., Polman, A.: Fabry-Pérot resonators for surface plasmon polaritons probed by cathodoluminescence. Appl. Phys. Lett. **94**, 183104 (2009). ISSN: 0003-6951, 1077-3118
48. Santoro, F., et al.: Interfacing electrogenic cells with 3D nanoelectrodes: position, shape, and size matter. ACS Nano **8**, 6713–6723 (2014)
49. Yetisen, A.K., et al.: Art on the nanoscale and beyond. Adv. Mater. **28**(9): 1724–1742 (2015)
50. Ferry, V.E., Polman, A., Atwater, H.A.: Modeling light trapping in nanostructured solar cells. ACS Nano **5**, 10055–10064 (2011)
51. Llordés, A., Garcia, G., Gazquez, J., Milliron, D.J.: Tunable near-infrared and visible-light transmittance in nanocrystal-in-glass composites. Nature **500**, 323–326 (2013)
52. Stockman, M.I.: Nanoplasmonics: the physics behind the applications. Phys. Today **64**, 39–44 (2011)
53. Pendry, J.B., Schurig, D., Smith, D.R.: Controlling electromagnetic fields. Science **312**, 1780–1782 (2006)
54. Ratanatawanate, C., Tao, Y., Balkus Jr., K.J.: Photocatalytic activity of PbS quantum dot/TiO2 nanotube composites. J. Phys. Chem. C **113**, 10755–10760 (2009)
55. Li, C., et al.: Grain-boundary-enhanced carrier collection in CdTe solar cells. Phys. Rev. Lett. **112**, 156103 (2014)
56. Ehrenreich, H., Cohen, M.H.: Self-consistent field approach to the many-electron problem. Phys. Rev. **115**, 786–790 (1959)
57. Adler, S.L.: Quantum theory of the dielectric constant in real solids. Phys. Rev. **126**, 413–420 (1962)
58. Wiser, N.: Dielectric constant with local field effects included. Phys. Rev. **129**, 62–69 (1963)
59. Koch, S.W., Kira, M., Khitrova, G., Gibbs, H.: Semiconductor excitons in new light. Nat. Mater. **5**, 523–531 (2006)
60. Ponseca Jr., C.S., et al.: Organometal halide perovskite solar cell materials rationalized: ultrafast charge generation, high and microsecond-long balanced mobilities, and slow recombination. J. Am. Chem. Soc. **136**, 5189–5192 (2014)
61. Schmidt-Mende, L., MacManus-Driscoll, J.L.: Zno-nanostructures, defects, and devices. Mater. Today **10**, 40–48 (2007)
62. Bhattacharya, A., Bhattacharya, S., Das, G.: Band gap engineering by functionalization of BN sheet. Phys. Rev. B **85**, 035415 (2012)
63. Poplawsky, J.D., Nishikawa, A., Fujiwara, Y., Dierolf, V.: Defect roles in the excitation of Eu ions in Eu: GaN. Opt. Exp. **21**, 30633–30641 (2013)
64. Drude, P.: Zur elektronentheorie der metalle. Ann. Phys. **306**, 566–613 (1900)
65. MacDonald, K.F., Sámson, Z.L., Stockman, M.I., Zheludev, N.I.: Ultrafast active plasmonics. Nat. Photon. **3**, 55–58 (2009). ISSN: 1749-4885
66. Kuttge, M., et al.: Local density of states, spectrum, and far-field interference of surface plasmon polaritons probed by cathodoluminescence. Phys. Rev. B **79**, 113405 (2009)
67. Zayats, A.V., Smolyaninov, I.I., Maradudin, A.A.: Nano-optics of surface plasmon polaritons. Phys. Rep. **408**, 131–314 (2005)
68. Bozhevolnyi, S.I., Volkov, V.S., Devaux, E., Laluet, J.-Y., Ebbesen, T.W.: Channel plasmon subwavelength waveguide components including interferometers and ring resonators. Nature **440**, 508–511 (2006)
69. Walther, R., et al.: Coupling of Surface-plasmon-polariton-hybridized cavity modes between submicron slits in a thin gold film. ACS Photon. **3**, 836–843 (2016)

70. Gong, S., et al.: Electron beam excitation of surface plasmon polaritons. Opt. Exp. **22**, 19252–19261 (2014)
71. Yamamoto, N., Suzuki, T.: Conversion of surface plasmon polaritons to light by a surface step. Appl. Phys. Lett. **93**, 093114 (2008)
72. Schoen, D.T., Atre, A.C., García-Etxarri, A., Dionne, J.A., Brongersma, M.L.: Probing complex reflection coefficients in one-dimensional surface plasmon polariton waveguides and cavities using STEM EELS. Nano Lett. **15**, 120–126 (2014)
73. Mie, G.: Beiträge zur Optik trüber Medien, speziell kolloidaler Metallösungen. Ann. Phys. **330**, 377–445 (1908)
74. Kreibig, U., in collab. with Vollmer, M.: Optical Properties of Metal Clusters. Springer Series in Materials Science, vol. 25, 532 pp. Springer, Berlin, New York (1995). ISBN: 0-387-57836-6
75. Pellegrini, G., Mattei, G., Bello, V., Mazzoldi, P.: Interacting metal nanoparticles: optical properties from nanoparticle dimers to core-satellite systems. Mater. Sci. Eng. C **27**, 1347–1350 (2007)
76. Stockman, M.I.: Nanofocusing of optical energy in tapered plasmonic waveguides. Phys. Rev. Lett. **93**, 137404 (2004)
77. Willets, K.A., Van Duyne, R.P.: Localized surface plasmon resonance spectroscopy and sensing. Annu. Rev. Phys. Chem. **58**, 267–297 (2007)
78. Boltasseva, A., Atwater, H.A.: Low-loss plasmonic metamaterials. Science **331**, 290–291 (2011)
79. Brongersma, M.L., Shalaev, V.M.: The case for plasmonics. Science **328**, 440–441 (2010)

Chapter 2
Tools and Techniques

Chapter 1 showed many of the nanoscale phenomena that have allowed great progress to be made in nanotechnology in recent years. The sophistication of these nanostructures makes high-precision analysis at the nanoscale an absolute must, and scientists have risen to the occasion for such analyses from both theoretical and experimental vantages. This chapter will delve into the methods used in this dissertation, provide the fundamental basics, and go into more detail about the important aspects of both the theoretical and experimental tools for nanoscale analysis and beyond.

2.1 Density Functional Theory: Quantum Mechanics for Complex Systems

One of the most prominent techniques for analysis at the nanoscale is density functional theory (DFT). DFT represents a methodology to treat a system with a large number of charged particles by determining its electron density as opposed to solving the many-body Schödinger equation [1, 2].

The many-body Schrödinger equation is expressed in terms of Coulomb potential between all the charged particles in the solid:

$$\widehat{H}_{\text{MB}} = \frac{-\hbar^2}{2} \left(\sum_{i=1}^{N_e} \frac{\nabla_{\mathbf{r}_i}^2}{m_0} + \sum_{\alpha=1}^{N_i} \frac{\nabla_{\mathbf{R}_\alpha}^2}{M_\alpha} \right) + \sum_{i=1}^{N_e} \sum_{j=i+1}^{N_e} \frac{e^2}{|\mathbf{r}_i - \mathbf{r}_j|} + \sum_{i=1}^{N_e} \sum_{\alpha=1}^{N_i} \frac{Z_\alpha e^2}{|\mathbf{r}_i - \mathbf{R}_\alpha|}$$

$$+ \sum_{\alpha=1}^{N_i} \sum_{\beta=\alpha+1}^{N_i} \frac{Z_\alpha Z_\beta e^2}{|\mathbf{R}_\alpha - \mathbf{R}_\beta|} \tag{2.1}$$

$$\Psi_{\text{MB}} = \psi(\mathbf{r}_1 \ldots \mathbf{r}_{N_e}, \mathbf{R}_1 \ldots \mathbf{R}_{N_i}) \tag{2.2}$$

where N_e and N_i are the total numbers electrons and ions in the system, $\mathbf{r}_n$ and $\mathbf{R}_\alpha$ are the vectors from the arbitrary origin of the system to the ith electron or αth ion, and Z_α and M_α are the charge and mass of the αth ion. While for smaller atoms with simple electron configurations, such as H and He, the equation may be manageable, for more complex (and interesting) systems, Eq. (2.1) becomes highly complicated mathematically.

Many different techniques and approximations have been developed to solve the many-body Schrödinger equation for more complex systems from first-principles, but the complexity of the many-body wave function still scales unfavorably [3–8].

DFT provided a breakthrough by not solving the many-body Schrödinger equation at all and instead dealing directly with the electron density. The Hohenberg-Kohn theorems provided the framework for treating a complex system in terms of density [1]:

1. For an external potential influencing a system of electrons, $V_{\text{ext}}(\mathbf{r})$, there can only exist one corresponding electron density, $n(\mathbf{r})$.
2. The electron density that minimizes the energy of the total Hamiltonian is the true ground-state density of the system.

The one-to-one correlation of an external potential and the electron density is an extremely powerful statement, because the interaction Hamiltonian can be reduced to three terms: the kinetic energy of the electrons, $\widehat{T}$, the electron-electron interactions, $\widehat{V}_{\text{ee}}$, and the interaction of the electrons with an external potential, $\widehat{V}_{\text{ext}}$. The final term, $\widehat{V}_{\text{ext}}$, is based off of the Born-Oppenheimer approximation that the ions are static relative to the electrons, and hence the contribution of the ions to the Hamiltonian is essentially the equivalent of an external potential. Now, the first Hohenberg-Kohn theorem can be invoked to express the entire many-body Hamiltonian in terms of the electron density, and the total energy of the system becomes a functional of the density

$$E[n(\mathbf{r})] = \langle \Psi | \widehat{T} + \widehat{V}_{\text{ee}} + \widehat{V}_{\text{ext}} | \Psi \rangle = F[n(\mathbf{r})] + \int n(\mathbf{r}) V_{\text{ext}}(\mathbf{r}) d\mathbf{r} \qquad (2.3)$$

where $|\Psi\rangle$ is the many-body wave function and $F[n(\mathbf{r})] = \langle \Psi | \widehat{T} + \widehat{V}_{\text{ee}} | \Psi \rangle$.

The second theorem then shows that the genuine ground-state wave functions of the entire system can be computed variationally, by systematically minimizing the energy functional with respect to the electron density.

In order to actually perform DFT calculations, $F[n(\mathbf{r})]$ needs to be determined to avoid evaluating $\langle \Psi | \widehat{T} + \widehat{V}_{\text{ee}} | \Psi \rangle$ which still involves the many-body wave functions.

Kohn and Sham proposed a fictitious system of noninteracting charged particles that has the same charge density as the genuine system [2]. In the fictitious system, $F[n(r)]$ is written as the sum of the kinetic energy, the Coulomb potential of the electron density, and a term called the exchange-correlation energy which encapsulates the fermionic nature of the particles

$$F[n(\mathbf{r})] = T[n(\mathbf{r})] + \frac{1}{2} \int \frac{n(\mathbf{r})n(\mathbf{r}')}{|\mathbf{r} - \mathbf{r}'|} d\mathbf{r} d\mathbf{r}' + E_{\text{xc}}[n(\mathbf{r})] \qquad (2.4)$$

2.1 Density Functional Theory: Quantum Mechanics for Complex Systems

Now one can group all terms aside from the kinetic energy functional into a new term

$$\int n(\mathbf{r}) \left[V_{\text{ext}}(\mathbf{r}) + \frac{1}{2} \int \frac{n(\mathbf{r}')}{|\mathbf{r}-\mathbf{r}'|} d\mathbf{r}' + \frac{\delta E_{\text{xc}}[n(\mathbf{r})]}{\delta n(\mathbf{r})} \right] d\mathbf{r} = V_{\text{KS}}[n(\mathbf{r})] \quad (2.5)$$

The importance of this result is that one can now rewrite the Hamiltonian for the system in terms of many single-particle wave functions, as opposed to one many-body wave function

$$\left[\frac{-\hbar^2}{2m} \nabla_i^2 + V_{\text{KS}}(\mathbf{r}) \right] \psi_{\text{KS}_i}(\mathbf{r}) = E_i \psi_{\text{KS}_i}(\mathbf{r}) \quad (2.6)$$

$$n(\mathbf{r}) = \sum_i^N f_i |\psi_{\text{KS}_i}(\mathbf{r})|^2 \quad (2.7)$$

Up until this point, the equations are exact and only represent a reformulation of the many-body problem. However, the exchange-correlation functional is still unknown. In order to perform calculations on larger systems, one can formulate an approximation of $E_{\text{xc}}[n(\mathbf{r})]$ and then solve the above equations in a self-consistent and iterative manner.

The most common approximation is called the local density approximation (LDA), in which $E_{\text{xc}}[n(\mathbf{r})]$ is approximated at each position, $\mathbf{r}$, as if it were a homogenous electron gas with the charge density at $\mathbf{r}$. There are a wide range of other approximations as well, and by understanding the required level of accuracy for the specific type of calculation one wishes to perform, DFT can utilize these various approximations in order to perform quantum mechanical calculations on complex crystals with unit cells containing hundreds of atoms.

The strength of DFT is in the fact that it produces a value for the total energy of the system. From the energy the forces acting on the lattice ions can be derived, which allows for DFT to relax unit cells until the internal forces are approximately zero. Thus, structures can be generated that are accurate to within the approximation of $E_{\text{xc}}[n(\mathbf{r})]$. The ability to relax crystal structures allows DFT to treat localized effects, such as point defects and interfaces, and produce realistic estimates of their ionic and electronic properties.

2.1.1 Calculating Optical Properties with Density Functional Theory

In Chap. 1, it was stated that the fundamental quantity of interest for determining the optical response of the material is the dielectric function. Through the Kohn-Sham equations, Eqs. (2.6) and (2.7), there is now a way of determining it directly

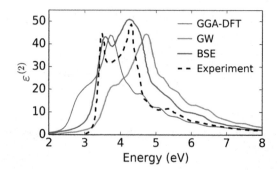

Fig. 2.1 Levels of accuracy in density functional theory. Here the calculation of ϵ_2 is shown via different levels of density functional theory calculations. The blue line shows the GGA-DFT calculation. The green and the red are the GW approximation and Bethe-Salpeter equation calculations, respectively, from [13]. The dashed black line shows the experimental values determined in [10]. From these plots it can be easily seen that the flavor of DFT can significantly affect the result

for an arbitrary crystal structure. The potential for this kind of atomistic-scale calculation of optical properties is profound and provides the opportunity to directly calculate the effect of localized features, such as interfaces and point defects, on the optical response of the material. However, the Kohn-Sham wave functions are only guaranteed to produce the correct *ground-state* properties, not the correct *excited-state* properties.

As a result there are some systematic errors in DFT calculations of the optical properties of materials, due to the inaccuracy of treating the excited states with the Kohn-Sham single-particle wave functions, alongside the embedded inaccuracy of approximating the exchange-correlation functional [9]. However, this has not slowed progress so much as provided a new field of study for the theorists in the community interested in the optical response of materials.

Figure 2.1 shows several different methods of calculating the imaginary part of the dielectric function, ϵ_2 of Si through DFT, as well as an experimental determination of ϵ_2 for comparison. The experimental curve is the dashed black line, taken from [10]. The blue curve shows ϵ_2 calculated within the generalized gradient approximation (GGA), which is similar to the LDA but also takes into account the gradient of the electron density [11]. Basic level DFT calculations such as LDA and GGA famously underestimate the bandgap of semiconductors (as seen here) and also fail to reproduce some other significant optical phenomena in materials.

To accommodate these issues, there are a few popular post-LDA/GGA processes that increase the accuracy of optical calculations. The *GW* approximation, which was developed independently from DFT at around the same time by L. Hedin [12], is one of the most prominent. In the *GW* approximation, the exchange-correlation potential is replaced with an energy-dependent self-energy that can be calculated at varying levels of approximation from the Green's function, G, and screened Coulomb potential, (W). *GW* calculations have shown to produce bandgaps that are fairly close the experimental values, as can be seen from the GW calculations from Rohlfing et al. [13], shown by the green line in Fig. 2.1. However, it can be seen that while the bandgap is more accurate in the GW approximation, it still does not match up to the experimental results in terms of the spectral shape.

Beyond the GW approximation, excitonic effects can be included through the Bethe-Salpeter equation; such calculations of the dielectric function of Si match both the bandgap and the spectral profile of the experimental values as shown by Rohlfing et al. [13], red line in Fig. 2.1.

Unfortunately, the availability of higher-accuracy calculations does not solve the problem, as the computational demand of these advanced methods is increased dramatically above standard DFT-level calculations, making the treatment of complex many-atom systems difficult. As a result, proper utilization of DFT involves choosing the level of the approximation effectively to maximize the information obtained from the calculation while keeping the computational workload manageable.

2.2 Finite-Difference Time-Domain: Electrodynamics for Nanostructures

DFT and other equivalent ab initio techniques are the standard for incorporating atomistic effects into theory. However, DFT calculations are only suitable for small systems or the unit cells of periodic bulk materials with size scales on the order of a few hundred atoms at maximum. Even small nanostructures, on the order of tens of nm in each dimension, correspond to thousands of atoms and are not viable for DFT.

However, as one escapes from the <10 nm size regime, quantum mechanical effects can be incorporated more loosely or ignored all together, and the optical response of the sample can be determined through solving Maxwell's equations, instead of Schrödinger's equation. Theories such as the ones put forth by Mie and Drude, discussed in the first chapter, provide excellent results for simple systems but are not suited for more complicated geometries. A methodology called finite-difference time-domain (FDTD) simulation was developed by Yee in 1966 [14], in which the system is divided into cells where the different components of the magnetic and electric fields are offset from each other in a periodic manner and solved sequentially.

The two Maxwell's equations that connect the electric and magnetic fields (Eqs. (1.2) and (1.4)) in curl form are

$$\frac{\partial B_\alpha}{\partial t} = \frac{\partial E_\beta}{\partial \gamma} - \frac{\partial E_\gamma}{\partial \beta} \tag{2.8}$$

$$\frac{\partial D_\alpha}{\partial t} = \frac{\partial H_\gamma}{\partial \beta} - \frac{\partial H_\beta}{\partial \gamma} - J_\alpha \tag{2.9}$$

where α, β, and γ represent Cartesian sets of x, y, and z. A complex structure can be broken down into a grid of Yee cells, where one can consider the finite volume of each cell in the grid, solving over a time-step function. Using this method, Maxwell's equations in Eqs. (2.8) and (2.9) are transformed to the following.

$$\frac{B_\alpha^{n+1/2}\left(i,j+\tfrac{1}{2},k+\tfrac{1}{2}\right) - B_\alpha^{n-1/2}\left(i,j+\tfrac{1}{2},k+\tfrac{1}{2}\right)}{\Delta t}$$
$$= \frac{E_\beta^n\left(i,j+\tfrac{1}{2},k+1\right) - E_\beta^n\left(i,j+\tfrac{1}{2},k\right)}{\Delta \gamma} - \frac{E_\gamma^n\left(i,j+1,k+\tfrac{1}{2}\right) - E_\gamma^n\left(i,j,k+\tfrac{1}{2}\right)}{\Delta \beta} \tag{2.10}$$

$$\frac{D_\alpha^n\left(i+\tfrac{1}{2},j,k\right) - D_\alpha^{n-1}\left(i+\tfrac{1}{2},j,k\right)}{\Delta t}$$
$$= \frac{H_\gamma^{n-1/2}\left(i+\tfrac{1}{2},j+\tfrac{1}{2},k\right) - H_\gamma^{n-1/2}\left(i+\tfrac{1}{2},j-\tfrac{1}{2},k\right)}{\Delta \beta}$$
$$- \frac{H_\beta^{n-1/2}\left(i+\tfrac{1}{2},j,k+\tfrac{1}{2}\right) - H_\beta^{n-1/2}\left(i+\tfrac{1}{2},j,k-\tfrac{1}{2}\right)}{\Delta \gamma} + J_\alpha^{n-1/2}\left(i+\tfrac{1}{2},j,k\right) \tag{2.11}$$

where n is the iteration of the time step and i, j, and k indicate the indices of the Yee cell for the step of the calculation. From these equations one can see the iterative process by which one solves Maxwell's equation on a per cell basis.

Finite-difference time-domain simulations can compute the optical response of any arbitrary geometry by breaking it up into a mesh of homogenous cells and solving Maxwell's equations in each one. Additionally, the fact that it is a time-domain calculation, as opposed to a frequency-domain calculation, allows for solving over a broad range of wavelengths in a single simulation simultaneously, as opposed to doing each wavelength individually. For the plasmonic systems that will be treated later in this dissertation, this is the theoretical method of choice.

2.3 Scanning Transmission Electron Microscopy: Ultrahigh Resolution Analysis

In order to experimentally observe the response of complex nanostructures, high-spatial resolution analytical techniques have become increasingly important in nanotechnology. Scanning transmission electron microscopy (STEM) offers one of the strongest combinations of analytical detection and spatial resolution available in modern technology and is the primary form of experimental technique used in this dissertation.

2.3 Scanning Transmission Electron Microscopy: Ultrahigh Resolution Analysis

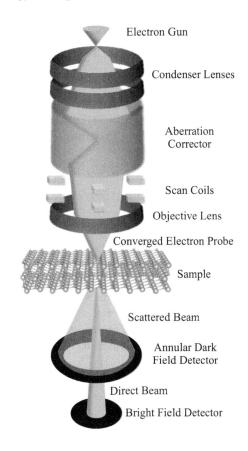

Fig. 2.2 Schematic of a scanning transmission electron microscope. An electron gun produces a beam of electrons that are formed into a coherent probe by the condenser lenses. The probe then passes through a series of multipole magnetic lenses that correct aberrations in the electron beam. The corrected probe then is rastered across the sample by scan coils, with the objective lens focusing the entirety of the probe to a point (up to < 0.5 Å diameter at best). The transmitted electrons are then collected in various detectors, including the annular dark field (ADF) detector for highly scattered electrons, and the bright field (BF) detector for elastically scattered electrons

The basic principle of STEM operation is demonstrated in Fig. 2.2. The source for the beam is called the electron gun. In my experiments I only use field emission guns (FEGs), which extract the electrons through a sharpened tungsten tip and then applying accelerating voltage to produce the high-energy electron probe. After the electrons are extracted from the gun, they pass through a series of magnetic lenses in order to form a converged, coherent, and small probe at the sample. First, the beam passes through the condenser lenses, which collect the extracted electrons and form a coherent electron beam. Once the beam is formed, the probe can be corrected; this is an important step for the resolution of the STEM and will be treated in more detail in the next section. Briefly, the corrector is a series of multipole magnetic lenses that can be used to remove spherical aberrations from the electron beam.

The corrected probe then passes through scan coils which are used to deflect the beam in a controlled manner to raster it across the sample, while the objective lens focuses the probe down to the smallest possible size. The size of the probe determines the spatial resolution of the microscope, and the smaller the probe size, the higher the resolution. Current aberration-corrected probes are capable of obtaining <50pm spatial resolution and precisions down to the single digit picometer range, as well as providing significant improvements to the depth resolution [15–20]. As the probe transmits through the sample, some electrons are scattered to high angles and are collected by the annular dark field (ADF) detector, while some are only elastically scattered and stay close to the optical axis and are collected by the bright field (BF) detector. The distinction between these two detectors is important and presents one of the key benefits of STEM, as will be further discussed later in this chapter. Additionally in Appendix A, the different microscopes that are used in this experiment are described in detail.

2.3.1 Correcting Aberrations in an Electron Probe

As stated earlier, the key to accessing the ultrahigh spatial resolutions in the electron microscope lies in aberration correction. Aberrations result from inhomogeneities in the magnetic lens, which means different electrons that pass through the lens at different points experience different magnetic fields, and are focused to different points. Figure 2.3 compares the ray trajectories of electrons going through a perfect lens (a), with the trajectories of electrons going through an aberrated lens (b). In the perfect lens, all rays converge to a single point, the focal point of the lens; however in the aberrated lens, electrons that interact with the edge of the lens are deflected

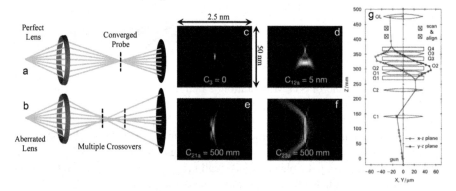

Fig. 2.3 Aberrations to electron wave fronts. When an electron beam interacts with a perfect aberration-free lens (**a**), all electrons are converged exactly to the focal point of the lens, allowing for infinite magnification there. In reality, all lenses have some form of aberration (**b**), which cause different ray trajectories to be deflected differently, which prevents the formation of a highly focused probe at a single point. (**c**)–(**f**) Profiles of electron probes in the x–z plane demonstrating the effects of spherical aberrations on electron probes (Figure from A. Lupini and N. de Jonge, "The Three-Dimensional Point Spread Function of Aberration-Corrected Scanning Transmission Electron Microscopy" *Microscopy and Microanalysis* **17**, 5, pp: 817–826 reproduced with permission.). (**g**) Schematic demonstrating the method of correcting spherical aberrations through quadrupole magnetic lenses (Reprinted from *Ultramicroscopy* **78**, O. Krivanek, N. Dellby, A. Lupini, "Towards sub-Åelectron beams", pp: 1–11, Copyright (1999), with permission from Elsevier.)

2.3 Scanning Transmission Electron Microscopy: Ultrahigh Resolution Analysis

more strongly than ones near the optic axis. The deviation causes a difference in the crossovers between the ray paths on the edge and those in the center, and as a result the crossovers of each ray occur at a different point in the z-axis. Since the electrons are not converged at a point, when the electron beam is rastered across the specimen, it samples a much larger area than just a converged point, which reduces the total spatial resolution.

Aberrations can take many forms, and the shape of the electron probe is essentially determined through the aberration function, given by the following expression:

$$\chi(\theta, \phi) = \sum_{n}\sum_{m=0}^{n+1} \left(C_{nma} \frac{\theta^{n+1}}{n+1} \cos(m\phi) + C_{nmb} \frac{\theta^{n+1}}{n+1} \sin(m\phi) \right), \quad (2.12)$$

where θ is the deflection from the optical axis, ϕ is the azimuthal angle in the x-y plane, and C_{nma} is the spherical aberration of order n, where $\frac{2\pi}{m}$ is the smallest angle that results in a phase shift (for cylindrically symmetric modes, m is not included in the index of the aberration), and a refers to whether the aberration affects the sin or cos term [21].

Figure 2.3c–f show the different effects of different types of aberrations on an electron probe from [22]. The figures show the profile of the electron probe intensity in the x-z plane and demonstrate the probe's ability to converge under different spherical aberrations. Figure 2.3c shows an aberration-corrected probe, which is converged and localized in the x-z plane. However, when aberrations are added (C_{12a} in d, C_{21a} in e, and C_{23a} in f), it can be seen that the probe becomes irregular, and multiple problematic effects arise including multiple local maxima, intensity delocalized throughout the z-dimension, and profile x plane spread out across a larger area. It is straightforward to see how imaging with such probes would make imaging sub-nm features difficult.

Spherical aberration correction was first predicted by Scherzer in 1949. He noted that while the aberrations induced by round lenses were always positive, they could be cancelled out by inducing negative aberrations with high-order, multipole magnetic lenses [23]. The aberration-corrected microscopy performed in this work is all done on correctors developed in the early 2000s based on Scherzer's design; a schematic of which from [21] is shown in Fig. 2.3g. Now, aberration correction is standard in high-end STEMs and TEMs and sub-Angstrom resolution is routinely achieved [24].

2.3.2 Bright Field and Dark Field in the STEM

When comparing electron microscopy to optical microscopy, the closest analogue is TEM. In TEM, the sample is most often illuminated with a beam, where the electrons travel nearly parallel to the optical axis and illuminate a large area.

The electrons then travel through projector lenses, to reform an image of the sample at the CCD detector, similar to the way an optical microscope reforms the image of the sample at the eyepiece. However, this presents some problems [25]. As the coherent beam of electrons pass through a sample, slight changes are made to the phase of each electron, along with the electrons undergoing elastic scattering. Diffraction and phase contrast provide deep information about the sample; however, the convolution of multiple different contrast effects in TEM makes it difficult to directly interpret the image and makes rigorous quantitative understanding of TEM images a significant challenge.

As a result, STEM has gained popularity with respect to TEM in recent years due to the availability of high-angle annular dark-field (HAADF) imaging. In a STEM, after transmitting through the sample, most electrons have not been scattered significantly and are within a few milliradians of the optic axis. Such electrons can be collected by the BF detector, but electrons collected here still possess all of the phase contrast observed in a standard TEM. However, many electrons undergo Rutherford scattering when interacting with the sample and are deflected out to higher angles. By placing an annular detector with a large inner angle to collect these electrons, one can form an incoherent image of the sample based only on the scattering potential. The result is called Z-contrast imaging, since the intensity in the image is approximately proportional to the square of the Z value of the probe column, where Z is the total sum of the atomic numbers of the atoms at the probe position.

Figure 2.4 shows an image of a Ge FinFET acquired with both the BF and HAADF detectors. The HAADF (a) shows a relatively uniform contrast across

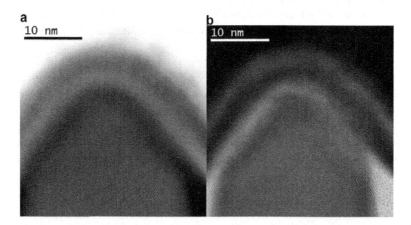

Fig. 2.4 Bright field vs. dark field imaging. Images of Ge FinFET are acquired with both high-angle annular dark-field (HAADF) detection (**a**) and BF detection (**b**). Due to phase contrast in the BF image, a strain/defect layer can be seen in the BF image, but the origin and nature of the layer cannot be directly inferred from the image. The HAADF image on the other hand is directly interpretable and reveals that the overall composition throughout the channel is relatively unchanged

the entire channel, while the BF (b) image shows a dark contrast layer along the edge corresponding to a strain or defect layer on the outer edge of the channel. The BF image shows that there is some type of structural difference between the outer edge of the channel and the inside but provides no direct information on what the difference might be. The HAADF image, on the other hand, does not show any differences in contrast between the outer edge and the inner portion of the channel. Since the HAADF is only based off of Z-contrast, it is directly interpretable and shows that the composition of the channel is for the most part unchanged. The disparity between BF and HAADF indicates that the dark contrast layer in the BF image is due to strain or some other effect that does not change the stoichiometry of the channel. This example demonstrates the versatility of STEM and how different imaging modes can be used to include or exclude different effects depending on the desired result of the user.

2.4 Electron-Beam Spectroscopies for Nanoscale Optical Properties

As Sect. 2.3.2 demonstrated, one of the primary benefits of STEM is the ability to image by correlating the signal from an analytical detector with the position of a converged probe. In addition to the imaging capabilities of the BF and ADF detectors, spectroscopy is also commonly performed in STEM. The two techniques that will be focused on in this dissertation are electron energy loss spectroscopy (EELS) and cathodoluminescence (CL).

Figure 2.5 shows a schematic for both spectroscopies and how they are acquired in the electron microscope. For EELS the direct transmitted electron beam is captured and then sent through a magnetic prism. Electrons with greater energy are deflected more by the magnetic field, as a result all the electrons are dispersed by how much energy they lost during transmission through the sample. This energy lost during transmission corresponds to energy transferred into the sample, and as a result EELS is essentially a direct measure of all excitations induced by the electron beam in the sample.

Conversely, CL only uses electrons for excitation, not detection. The beam electrons induce excitations in the sample, but the signal comes from conduction band electrons returning to ground state via radiative decay. As photons are emitted from the sample, they are collected by a parabolic mirror, which sits directly above the sample. A small hole is drilled through the top of the mirror to allow the electron beam to pass through, but the size is negligible compared to the rest of the mirror. The reflected light travels out of a port on the side of the microscope or is fiber-coupled out of the machine where it is collected as CL [26].

The two techniques can probe similar phenomena (such as surface plasmons), but by using different signals for detection, they become highly complementary. EELS, by using the transmitted electrons, measures excitation. As a result it can

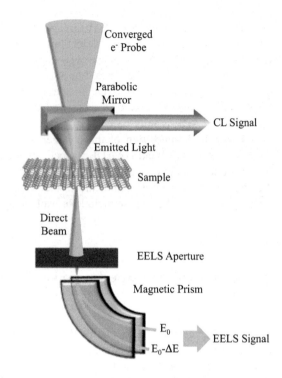

Fig. 2.5 Spectroscopy in the electron microscope. Schematics for the operation of EELS and CL are shown. The converged probe is rastered across the sample. From this probe one can collect the transmitted electrons, disperse them by energy lost during transmission with a magnetic prism, and collect them for EELS. Or they can examine the radiative decay from beam-induced excitation, reflect the light out of a port in the side of the microscope using a parabolic mirror, and collect the photons as CL. The two techniques are highly complementary due to the fact that EELS is a measure of excitation, and CL is a measure of the decay of those same samples

miss optical phenomena such as excitons and defect peaks, which have low beam interaction cross sections, but play a large role in the emission spectrum. CL, by using the emitted light, measures radiative decay in the sample. So luminescent phenomena such as excitons and defect states are easily observed, but non-radiative excitations (such as bulk plasmons) cannot be detected. More on the complementarity between the two will be discussed later in Chap. 5.

2.4.1 Electron Beam Interactions with Materials

In Sect. 1.3.1, the energy-loss function was introduced as the response of a solid to an incoming fast electron. Since STEM-based spectroscopies utilize a converged probe of fast electrons, it becomes appropriate to discuss the loss function in greater detail. Taking a step back from the loss function introduced earlier (Eq. (1.8)), it is useful to first try to intuitively consider a fast electron travelling through some dielectric medium and how much energy the electron imparts to the medium during transmission. By considering two different ways of considering the energy loss, one can develop an intuitive physical picture of how optical excitations in the electron microscope are detected [27, 28]. The first is based on the work, W, done on the electron moving with velocity, v, by the induced field, $\mathbf{E}_{ind}$

$$W = -e \int_{-\infty}^{\infty} \mathbf{v} \cdot \mathbf{E}_{ind} dt \qquad (2.13)$$

2.4 Electron-Beam Spectroscopies for Nanoscale Optical Properties

Equation (2.13) provides a real-space picture for how a fast electron loses energy. However, one can also consider a hypothetical energy loss probability, $\Gamma(\omega)$, that simply provides the probability of the electron losing $\hbar\omega$ of energy during transmission

$$W = \int_0^\infty \hbar\omega \Gamma(\omega) d\omega \tag{2.14}$$

Both equations measure the total energy loss of the electron and should be equivalent, which allows for one to write the loss probability in terms of the Fourier transform of the induced field [28]

$$\Gamma(\omega) = \frac{-e}{\pi\hbar\omega} \int_{-\infty}^\infty \text{Re}\left[e^{-i\omega t} \mathbf{v} \cdot \mathbf{E}_{\text{ind}}\right] dt. \tag{2.15}$$

The induced field can then be expressed as the product of the Green's function of the electron, and Eq. (2.15) can be reduced to

$$\Gamma(\omega) = \frac{4e^2}{\hbar} \text{Im}\left[\hat{\mathbf{v}} \cdot \mathbf{G}(q, \mathbf{R}_0, \omega) \cdot \hat{\mathbf{v}}\right], \tag{2.16}$$

where q is the wavevector of the electron and R_0 is the vector describing the position of the electron on the plane normal to $\mathbf{v}$. To place a physical meaning on this quantity, the loss probability can be related to the local density of states (LDOS) which can also be expressed in terms of a trace of the Green's function [29])

$$\rho_{\hat{\mathbf{v}}}(\mathbf{r}, \omega) = -\frac{2\omega}{\pi} \text{Im}\left[\hat{\mathbf{v}} \cdot \mathbf{G}(\mathbf{r}, \omega) \cdot \hat{\mathbf{v}}\right] \tag{2.17}$$

So the loss probability truly is a measure of the LDOS on the path of the beam, with wavevector q and frequency ω

$$\Gamma(q, \omega) = -\frac{2\pi e^2}{\hbar\omega} \rho_{\hat{\mathbf{v}}}(q, \mathbf{R}_0, \omega) \tag{2.18}$$

Equation (2.18) provides a mathematical formula to a physically intuitive picture of the energy loss of an electron. As the electron travels through some medium, there is some probability that it loses energy by creating excitations in the sample. As a result, the electron beam is an excellent tool for studying the optical properties of materials because of its ability to probe optical transitions at the local level.

2.4.2 Electron Energy Loss Spectroscopy

In the world of STEM spectroscopy, there is perhaps no technique more important than EELS, due to its ability to provide extremely localized spectroscopic data, while simultaneously acquiring HAADF signal [30–32]. Not only does EELS bring spectroscopy down to the atomic level, but it also provides two distinct modes of analysis that are each highly important and highly utilized in the field of materials science.

Core-Loss EELS One of the advantages of electron microscopy over other techniques is the ability to impart the electron probe with extremely high energies. The advantage of this lies in the interaction of the high-energy electrons with heavier atoms. Heavy atoms have deep-core states that are extremely stable. However, STEMs are typically operated between 60 and 300 keV, meaning the beam electrons can easily impart the needed energy into a deep-core electron to excite it directly into an unoccupied state in the conduction band. Transitions from each different electron level have their own signature energies, i.e., the Si-K edge (from the 1s core level) has its energy loss at 1839 eV, while L-edge (from the shallower 2p levels) are at 99 eV. The importance of the technique lies in the fact that the EELS edges are unique to each element, allowing for one to sample the chemical composition, beyond the capabilities of HAADF imaging, of a sample with the precision of the electron probe.

Figure 2.6a shows the core-loss EEL spectrum from an iron oxide nanoparticle on a gold plasmonic nanotriangle (a sample that will be discussed more in depth in Sect. 4.2). The plot shows the O K-edge at 538 eV and the Fe L_3- and L_2-edges at 708 eV and 721 eV, respectively, indicating that the nanoparticles are indeed iron oxide. A closer look at the edges in our core-loss spectrum reveals a great deal of fine structure to the EELS edges. The exact location of different features in the fine structure contains information based off of the valency and composition of the material at a local level and can be used to experimentally study defects with high precision [33–36].

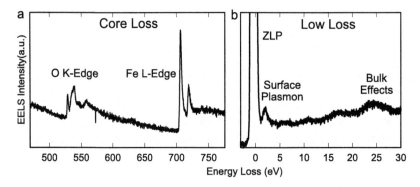

Fig. 2.6 Modes of electron energy loss spectroscopy. (**a**) The core-loss EEL spectrum, which contains structural and chemical information about the sample. (**b**) The low-loss region. The sharpest feature is the zero loss peak (ZLP), corresponding to elastic scattering in the sample. The peaks to the far right are bulk effects such as interband transitions and bulk plasmons, which give dielectric and thickness information about the sample, and surface plasmons can also be observed near the ZLP

Low-Loss EELS Figure 2.6b shows the regime called low loss for the same sample as Fig. 2.6a. Here the dominant features are bulk effects, such as bulk plasmons and interband transitions, and the most prominent feature is the zero loss peak (ZLP) which contains the counts of the elastically scattered electrons that only loosely interact with the sample during transmission. The strength of the energy loss due to volume effects is proportional to thickness, which makes analysis over thick samples very difficult in EELS; however, as will be discussed in Sect. 5.3, careful analysis of the bulk effects and the ZLP can still provide valuable information about the sample.

For the purposes of optical analyses, the regime directly next to the ZLP is of interest, as this is the region where surface plasmons (as well as other optical phenomena such as bandgaps) can be observed, as seen in Fig. 2.6c. Along the edges of the sample, the free surface charge of the metal creates a high LDOS to be sampled by the beam electrons. The result is strong excitation of the LSPRs of the nanoparticle at the points of the highest electric field enhancement of the plasmon mode. The plasmon modes are delocalized effects, so imaging to the precision that is achievable in core-loss EELS is inherently impossible. Still, performing EEL spectroscopy with high dispersions and minimized ZLPs, plasmon modes can be observed and mapped with precision.

2.4.3 Cathodoluminescence

The other technique employed heavily in this dissertation is CL. There is a rich history of research in CL, but a large portion of it has been done on the SEM [37–41]. However, recently CL in a STEM has become a more popular option [26, 42–45]. Traditionally, in SEM-CL the beam current is ramped up to put a higher dose of energy into the sample and get more signal, and then the accelerating voltage is reduced to try to localize the signal. However, as can be seen from the Monte Carlo simulation of 3 keV electrons being injected into 100 nm of ZnO, the SEM beam still samples a huge interaction volume that greatly reduces the spatial resolution, (Fig. 2.7a).

STEM-CL uses much higher-energy electrons that in turn sample a much smaller area, and the area of excitation is much more tightly localized to beam column than SEM-CL as can be seen in Fig. 2.7b, for the Monte Carlo simulation of high-energy 60- keV electrons. The difficulty with STEM-CL is that as a result of this localized excitation, the total beam-sample interaction is greatly reduced from the 3- keV case, and the total CL signal is much lower. This couples with the fact that CL acquisition is a lossy process, and many of the processes of interest for CL are only weakly luminescent to begin with. So at a fundamental level, the only real difference between SEM- and STEM-CL is the accelerating voltage, and the two techniques are only separated by the availability of low-keV accelerating voltages in SEMs and high-accelerating voltages in STEMs; the advantage of STEM-CL really comes with the ability to correlate the STEM-CL images with other analytical detectors in STEM such as EELS and ADF imaging.

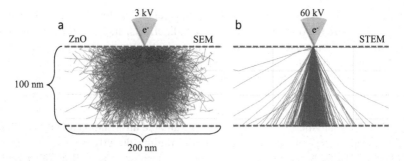

Fig. 2.7 STEM-CL vs. SEM-CL. Monte Carlo simulations of the interaction volume in 100 nm of ZnO for (**a**) SEM-CL with an accelerating voltage of 3 kV, and (**b**) STEM-CL with an accelerating voltage of 60 kV. While SEM-CL has the advantage of higher collected signal, the higher accelerating voltages of STEM-CL allows for higher spatial resolution in optical experiments

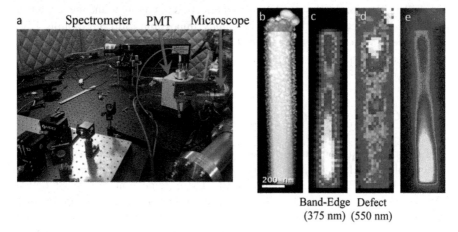

Fig. 2.8 Cathodoluminescence in a STEM. (**a**) The configuration for the CL experiments in this dissertation. The set-up allows for a few different modes of CL: spectroscopy and PMT-CL imaging. (**b**) shows a HAADF image of ZnO nanowire with Ag nanoparticles on the surface. (**c**) and (**d**) show spectrum images (SI) of different features acquired through spectroscopy, (**c**) being the band-edge exciton and (**d**) being the defect luminescence. (**e**) shows a STEM image taken with a PMT directly in front of the CL port, which allows for high-efficiency, but nonspectrally resolved imaging

The benefits of STEM-CL with respect to EELS are, firstly, the ability to isolate luminescent excitations and phenomena from non-radiative ones which has already been discussed, but also there is the versatility that results from utilizing a photon signal through electron excitation. Figure 2.8 shows the configuration of the CL microscope used for the experiments in this dissertation. On the right hand side of Fig. 2.8a sits the microscope (a VG-HB601). There is a port on the side of the microscope that allows the light to leave the column and enter an optics set-up. From there the beam can be directed to a spectrometer. Figure 2.8b shows a HAADF image of ZnO nanowire decorated with Ag nanoparticles, while Fig. 2.8c, d shows intensity maps of the different optical features of the sample, called spectrum images (SI): (c) the band-edge exciton emission SI and (d) the defect emission SI.

CL can also directly image the luminescent intensity of the sample, by placing a high-efficiency photomultiplier tube (PMT) detector directly in front of the CL port on the microscope. The PMT collects the signal and gives a direct readout of the total current, which corresponds directly to the total luminescence in the sample. PMT-CL imaging allows for high signals but at the cost of spectral resolution. Figure 2.8e shows the PMT-CL image of the nanowire from b–d. The higher resolution is easy to see, but it is not possible to distinguish between the defect and band-edge luminescence. By choosing the right technique at the right time, the savvy materials scientist can maximize the aspects of CL needed for each part of the experiment by utilizing these different set-ups effectively. Additionally, it can be seen in Fig. 2.8a that the entire set-up is on an optics table. This enables the possibility of future advanced experiments combining photonics and electron microscopy, which will be discussed further in Chap. 6.

Spectral resolution must also be considered a significant advantage of CL. In EELS the spectral resolution is capped by instrumental limitations, as a standard cold FEG typically possesses no better than 300- meV energy resolution [46]. Monochromation can greatly improve the energy resolution of the instrument, but even the best instruments worldwide are maxed out at 8- meV resolutions, and most do not even get that far [46–48]. The spectral resolution of CL, however, is only limited by one piece of equipment, the diffraction grating, which while wavelength dependent usually has excellent spectral resolution, <1 meV.

As a result, in most CL experiments, the spectral resolution is almost negligible because the actual linewidth of the optical features is significantly broader than the spectral resolution limit of the grating. Figure 2.9a–c show different segments of the same spectra from GaN doped with Eu. The sample is extremely bright, and the dopant atoms sitting in different positions result in a huge number of different emission peaks [49]. The main peaks have widths on the order of 10–15 meV, but the smaller satellite peaks in Fig. 2.9c have full-width half-maximums (FWHMs) of <5 meV, and peaks separated by less than 5 meV are easily observed.

There is another important distinction to make as the limits of CL are discussed. In terms of spatial resolution, it turns out that the interaction volume does not end up being the limiting factor in most STEM-CL experiments, because charge carrier diffusion ends up being the primary culprit. Figure 2.9d shows a BF-STEM image of a lacey carbon TEM grid with a sparse distribution of nanoparticles on the mesh, while Fig. 2.9e shows the simultaneously acquired PMT-CL image. The important thing to note from this comparison is that far away from the copper grid, there is some faint luminescence emanating from the lacey carbon. However, that luminescence disappears as the beam draws closer to the copper grid, and even as far as a few microns away, the luminescence is quenched. The reason is that lacey carbon is conduction, so charge carriers excited by the beam can diffuse throughout the lacey carbon, and near the copper grid the carriers find a non-radiative path to ground, while far away there is a high enough rate of radiative decay for luminescence to be detected.

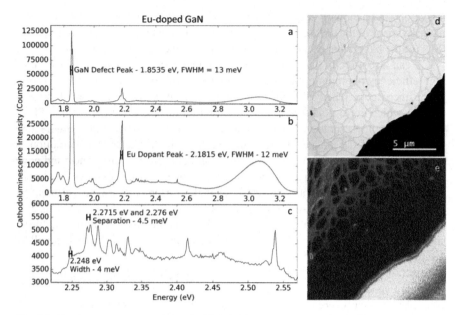

Fig. 2.9 Spatial and energy resolution in STEM-CL. (**a**)–(**c**) show a single CL spectrum from a bright Eu-doped GaN quantum well at different scales and show the various peaks that the dopant atoms create in the GaN bandgap. Features as small as 4 meV can be observed since the energy resolution of STEM-CL is less than the linewidth of the features. (**d**) An ADF image of a lacey carbon film near a copper grid. (**e**) A simultaneously acquired PMT-CL image of the same grid. It can be seen that the carbon is luminescent away from the copper, but not close to it. The difference is due to charge carriers near the copper grid diffusing and recombining non-radiatively at the copper grid, showing that the spatial resolution of STEM-CL is limited by charge carrier diffusion in the sample, not the probe size

Together CL and EELS present two excellent techniques for attacking nanoscale optical complexity from a variety of different angles and to really push our understanding of highly complicated nanostructures to the next level.

References

1. Hohenberg, P., Kohn, W.: Inhomogeneous electron gas. Phys. Rev. **136**, B864–B871 (1964)
2. Kohn, W., Sham, L.J.: Self-consistent equations including exchange and correlation effects. Phys. Rev. **140**, A1133–A1138 (1965)
3. Thomas, L.H.: The calculation of atomic fields. Math. Proc. Camb. Philos. Soc. **23**, 542–548 (1927)
4. Fermi, E.: Un metodo statistico per la determinazione di alcune priorieta dell'atome. Rend. Accad. Naz. Lincei **6**, 32 (1927)
5. Hartree, D.R.: The wave mechanics of an atom with a non-Coulomb central field. Part I. Theory and methods. Math. Proc. Camb. Philos. Soc. **24**, 89–110 (1928)

References

6. Hartree, D.R.: The wave mechanics of an atom with a non-Coulomb central field. Part II. Some results and discussion. Math. Proc. Camb. Philos. Soc. **24**, 111–132 (1928)
7. Fock, V.: Näherungsmethode zur Lösung des quantenmechanischen Mehrkörperproblems. Z. Phys. **61**, 126–148 (1930)
8. Slater, J.C.: A simplification of the Hartree-Fock method. Phys. Rev. **81**, 385 (1951)
9. Wang, C., Pickett, W.: Density-functional theory of excitation spectra of semiconductors: application to Si. Phys. Rev. Lett. **51**, 597 (1983)
10. Aspnes, D., Theeten, J.: Spectroscopic analysis of the interface between Si and its thermally grown oxide. J. Electrochem. Soc. **127**, 1359–1365 (1980)
11. Perdew, J.P., Burke, K., Ernzerhof, M.: Generalized gradient approximation made simple. Phys. Rev. Lett. **77**, 3865–3868 (1996)
12. Hedin, L.: New method for calculating the one-particle Green's function with application to the electron-gas problem. Phys. Rev. **139**, A796 (1965)
13. Rohlfing, M., Louie, S.G.: Electron-hole excitations and optical spectra from first principles. Phys. Rev. B **62**, 4927–4944 (2000)
14. Yee, K.S., et al.: Numerical solution of initial boundary value problems involving Maxwell's equations in isotropic media. IEEE Trans. Antennas Propag. **14**, 302–307 (1966)
15. Bals, S., Van Aert, S., Van Tendeloo, G., Ávila-Brande, D.: Statistical estimation of atomic positions from exit wave reconstruction with a precision in the picometer range. Phys. Rev. Lett. **96**, 096106 (2006)
16. Borisevich, A.Y., Lupini, A.R., Pennycook, S.J.: Depth sectioning with the aberration-corrected scanning transmission electron microscope. Proc. Natl. Acad. Sci. USA **103**, 3044–3048 (2006)
17. Kisielowski, C., et al.: Detection of single atoms and buried defects in three dimensions by aberration-corrected electron microscope with 0.5-Å information limit. Microsc. Microanal. **14**, 469–477 (2008)
18. Muller, D.A.: Structure and bonding at the atomic scale by scanning transmission electron microscopy. Nat. Mater. **8**, 263–270 (2009)
19. Lupini, A.R., et al.: Characterizing the two-and three-dimensional resolution of an improved aberration-corrected STEM. Microsc. Microanal. **15**, 441–453 (2009)
20. Yankovich, A.B., et al.: Picometre-precision analysis of scanning transmission electron microscopy images of platinum nanocatalysts. Nat. Commun. **5**, 4155 (2014)
21. Krivanek, O., Dellby, N., Lupini, A.: Towards sub-Åelectron beams. Ultramicroscopy **78**, 1–11 (1999)
22. Lupini, A.R., De Jonge, N.: The three-dimensional point spread function of aberration-corrected scanning transmission electron microscopy. Microsc. Microanal. **17**, 817–826 (2011)
23. Scherzer, O.: The theoretical resolution limit of the electron microscope. J. Appl. Phys. **20**, 20–29 (1949)
24. Batson, P., Dellby, N., Krivanek, O.: Sub-ångstrom resolution using aberration corrected electron optics. Nature **418**, 617–620 (2002)
25. Williams, D.B., Carter, C.B.: Transmission Electron Microscopy, pp. 3–17. Springer, Berlin (1996)
26. Colliex, C., et al.: The STEM multi-signal approach: learning the most from your nano-object. Microsc. Anal. **6**, 33–42 (2012)
27. Ritchie, R.H.: Plasma losses by fast electrons in thin films. Phys. Rev. **106**, 874–881 (1957)
28. García de Abajo, F.J.: Optical excitations in electron microscopy. Rev. Mod. Phys. **82**, 209–275 (2010)
29. Fussell, D., McPhedran, R., De Sterke, C.M.: Three-dimensional Green's tensor, local density of states, and spontaneous emission in finite two-dimensional photonic crystals composed of cylinders. Phys. Rev. E **70**, 066608 (2004)
30. Lupini, A., Pennycook, S.: Localization in elastic and inelastic scattering. Ultramicroscopy **96**, 313–322 (2003)
31. Kimoto, K., et al.: Element-selective imaging of atomic columns in a crystal using STEM and EELS. Nature **450**, 702–704 (2007)

32. Muller, D., et al.: Atomic-scale chemical imaging of composition and bonding by aberration-corrected microscopy. Science **319**, 1073–1076 (2008)
33. Browning, N., Wallis, D., Nellist, P., Pennycook, S.: EELS in the STEM: determination of materials properties on the atomic scale. Micron **28**, 333–348 (1997)
34. Arenal, R., et al.: Extending the analysis of EELS spectrum-imaging data, from elemental to bond mapping in complex nanostructures. Ultramicroscopy **109**, 32–38 (2008)
35. Suenaga, K., Koshino, M.: Atom-by-atom spectroscopy at graphene edge. Nature **468**, 1088–1090 (2010)
36. Zhou, W., et al.: Direct determination of the chemical bonding of individual impurities in graphene. Phys. Rev. Lett. **109**, 206803 (2012)
37. Rosner, S., Carr, E., Ludowise, M., Girolami, G., Erikson, H.: Correlation of cathodoluminescence inhomogeneity with microstructural defects in epitaxial GaN grown by metalorganic chemical-vapor deposition. Appl. Phys. Lett. **70**, 420–422 (1997)
38. Wu, X., Siu, G., Fu, C., Ong, H.: Photoluminescence and cathodoluminescence studies of stoichiometric and oxygen-deficient ZnO films. Appl. Phys. Lett. **78**, 2285–2287 (2001)
39. Saito, N., et al.: Low-temperature fabrication of light-emitting zinc oxide micropatterns using self-assembled monolayers. Adv. Mater. **14**, 418–421 (2002)
40. Chaturvedi, P., et al.: Imaging of plasmonic modes of silver nanoparticles using high-resolution cathodoluminescence spectroscopy. ACS Nano **3**, 2965–2974 (2009). ISSN: 1936-0851
41. Atre, A.C., et al.: Nanoscale optical tomography with cathodoluminescence spectroscopy. Nat Nano **10**, 429–436 (2015). ISSN: 1748-3387
42. Dadgar, A., et al.: Eliminating stacking faults in semi-polar GaN by AlN interlayers. Appl. Phys. Lett. **99**, 021905 (2011)
43. Kociak, M., et al.: Seeing and measuring in colours: electron microscopy and spectroscopies applied to nano-optics. C. R. Phys. **15**, 158–175 (2014). ISSN: 1631-0705. Seeing and measuring with electrons: transmission Electron Microscopy today and tomorrow
44. Hachtel, J., et al.: Probing plasmons in three dimensions by combining complementary spectroscopies in a scanning transmission electron microscope. Nanotechnology **27**, 155202 (2016)
45. Saito, H., Yamamoto, N.: Control of light emission by a plasmonic crystal cavity. Nano Lett. **15**, 5764–5769 (2015)
46. Krivanek, O.L., et al.: Vibrational spectroscopy in the electron microscope. Nature **514**, 209–212 (2014)
47. Lazar, S., Botton, G., Zandbergen, H.: Enhancement of resolution in coreless and low-loss spectroscopy in a monochromated microscope. Ultramicroscopy **106**, 1091–1103 (2006)
48. Mukai, M., et al.: Development of a monochromator for aberration-corrected scanning transmission electron microscopy. Microscopy **64**, 151–158 (2015)
49. Poplawsky, J.D., Nishikawa, A., Fujiwara, Y., Dierolf, V.: Defect roles in the excitation of Eu ions in Eu: GaN. Opt. Exp. **21**, 30633–30641 (2013)

Chapter 3
Extracting Interface Absorption Effects from First-Principles

The focus of this dissertation now switches from the introduction of analytical techniques to the actual application of those techniques in the study of nanoscale optical properties. In this chapter, first-principles DFT calculations of the dielectric function are used to determine the effects of an interface on the absorption of a multilayer heterostructure. These results and figures in this chapter are reproduced from [1] with permission from AIP Publishing.

For large scale structures, the optical properties can be determined straightforwardly from less computationally demanding methods. However, superlattices, nanocomposites, and other complex nanoscale structures can have dimensions where quantum mechanical effects become dominant and must be accounted for. Here I outline a method for quantitatively determining the interface contribution to light absorption from DFT in $NiSi_2$/Si superlattices, and demonstrate how it can be applied for the design multilayer heterostructures with selectable absorption.

3.1 Atomistic Interface Effects

Interfaces play an important role in determining the optical behavior of thin films [2, 3]. The effect of an interface on light absorption is of interest in laser optics [4, 5], optical sensors and detectors [6], and photovoltaics [7–9]. The electronic properties of an interface rapidly decay into bulk properties, as has been demonstrated in a variety of different areas [10–12]. As a result, in macroscopic systems where interfaces account for a negligibly small fraction of the solid, they are treated as merely boundaries between two absorbing media. The effects of interfaces, namely, scattering and reflections, are then treated accurately using classical electrodynamics and other semiclassical approximations [13–15]. On the other hand, for nanoscale systems such as thin-film multilayer heterostructures, interfaces can comprise a non-negligible volume fraction and play a significant role. It has been demonstrated that the distinct bonding at interfaces indeed gives rise to new features in the absorption spectra of heterostructures and must be included in order to model real device performance [16–20]. As technology moves towards the nanoscale, it is necessary to have accurate methods to quantitatively account for the absorptive effects of interfaces.

DFT calculations on supercells that contain the relevant interface and a few unit cells of the constituent materials can be used to study interface absorption effects. However, DFT calculations are subject to computational limitations so that the direct calculation of the optical response of a large and complex multilayer structure may not be possible. In order to treat such systems, macroscopic classical calculations are often invoked, but such calculations neglect interface bonding and are not suitable.

Many applications in modern nanotechnology involve large and highly complex composites of nanostructures [21–23]. Llordés et al. have recently demonstrated how changes in the bonding of ITO nanocrystals embedded in amorphous NbOx can be used to tune optical transmission and create "smart windows" [24]. By understanding the relationship between nanocrystal sparsity and the interface bonding effects on optical transmission, they rationally design systems with optical transmission in a selected range. In order to utilize these unique interface bonds in nanotechnology, methods to combine the versatility of macroscopic calculations with the accuracy of atomistic calculations are needed. To this end, I have developed a method to extract the atomistic interface effects on absorption in a quantitative manner. The interface effects are then contained in an independent term, such that they can be combined with macroscopic calculations to study large-scale systems with quantum mechanical detail.

3.1.1 Extracting Atomistic Interface Absorption Effects

In order to access the interface absorption directly, I consider the *absorbance*, defined as the product of the absorption coefficient of an absorbing medium and its thickness. The interface effects are extracted through calculating the difference in the absorbance between two different types of supercells: atomistic and macroscopic. The first supercell treats the interface atomistically, meaning the interface contains chemical bonds that are not present in either bulk material and generate unique absorption effects. In the other supercell, the interface is treated macroscopically, meaning both layers remain in their bulk phases and the interface only serves as a boundary.

First, DFT is used to calculate the absorption coefficient of the atomistic supercell and the bulk phases of $NiSi_2$ and Si (the method of calculating the absorption coefficient will be explained later). The atomistic absorbance uses the thickness (t_{atom} in Fig. 3.1) and absorption coefficient of the atomistic supercell. The macroscopic absorbance defines the thicknesses of each layer by dividing the thickness of the atomistic supercell up at the halfway point between the last unit cell of Si and the last unit cell of $NiSi_2$ ($t_{Macr-NiSi_2}$ and $t_{Macr-Si}$ in Fig. 3.1) and the DFT-

3.1 Atomistic Interface Effects

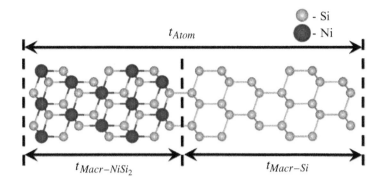

Fig. 3.1 Thicknesses for absorbance calculations. The DFT supercell is used to determine the thicknesses for both the macroscopic and atomistic cases. The total supercell length is the atomistic thickness, which is divided into the two macroscopic portions by the halfway point between the interface unit-cells of NiSi$_2$ and Si

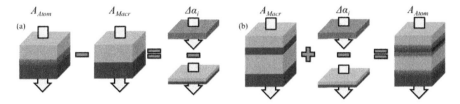

Fig. 3.2 Extracting interface absorption effects. (**a**) To quantitatively extract the effect of the interface, the difference between the atomistic absorbance, determined from absorption coefficients of relaxed superlattices containing interfaces, and the macroscopic absorbance, determined from the bulk absorption coefficients and total layer thicknesses of each material, is taken. (**b**) The resulting value is referred to as $\Delta\alpha_i$ and is an independent parameter that can then be combined with the macroscopic absorbance of different superlattices in order to determine the atomistic absorbance of the superlattices without the need for atomistic calculations

absorption coefficients of each bulk material. Since the thickness used to calculate the two absorbances are equivalent, the only difference between the two values comes from the differences in the absorption coefficient.

The macroscopic absorbance has only contributions from the bulk phases of each constituent material, while the atomistic absorbance has contributions from the bulk phases and the unique electronic structure in the interface regions. As a result, when the macroscopic absorbance is subtracted from the atomistic absorbance, shown schematically in Fig. 3.2a, the bulk absorbance common to both the macroscopic and atomistic treatments cancel out, leaving only the difference between an atomistic and macroscopic interface. The difference, $\Delta\alpha_i$, is an independent parameter that contains all of the information about the interface absorption effects. If multiple reflections have a negligible effect, pertinent $\Delta\alpha_i$ can be quantitatively added to the macroscopic absorbance of larger and more complex supercells, schematically demonstrated in Fig. 3.2b, to gain atomistic accuracy in absorption calculations for a supercell that cannot be treated directly with DFT (the limitations of the scheme with respect to multiple reflections will be addressed later).

3.1.2 $\Delta\alpha_i$ the Interface Absorbance Difference

To calculate $\Delta\alpha_i$, the absorption coefficient is needed, which in turn can be determined from the dielectric function (Eq. (1.6)). Ergo, the first step is to determine the dielectric functions of the atomistic supercell and each of the two bulk materials.

The DFT calculations of the dielectric functions are performed using the Vienna ab initio Software Package (VASP) [25, 26] within the projector-augmented wave framework [27] and using the Perdew-Burke-Ernzerhof exchange correlation functional in the GGA [28].

The imaginary part of the dielectric function is determined through the formulation put forth by Gajdoš et al. in [29] using the VASP code and the LOPTICS method where the imaginary part of the dielectric function is taken to be a weighted sum over direct transitions from the occupied valence band states to unoccupied conduction band states, and the real part is determined from a Kramers-Kronig Transformation as shown in Eqs. (3.1) and (3.2):

$$\epsilon_{\alpha\beta}^{(2)}(\omega) = \frac{4\pi^2 e^2}{\Omega} \tag{3.1}$$

$$\epsilon_{\alpha\beta}^{(1)}(\omega) = 1 + \frac{2}{\pi} P \int_0^\infty \frac{\epsilon_{\alpha\beta}^{(2)}(\omega')\omega'}{\omega'^2 - \omega^2} d\omega'. \tag{3.2}$$

For the bulk absorption coefficients in isotropic materials such as $NiSi_2$ and Si (cubic space group), all three components are the same. However, for the heterostructure calculations, the dielectric response for electric field normal to the interface (z-direction) differs from that of the in-plane directions (x-, y-directions). I consider light traveling normal to the interface, meaning all photons that propagate in the z-direction, which results in polarization in the x-y plane. The calculation of the dielectric function reflects this and has identical x and y components for the dielectric tensor; with a different spectrum in the z component, the relevant value is the x/y component. For all the calculations here, I use the x-component of the dielectric tensor, but using the y-component would produce an identical result.

These calculations require a large number of empty conduction bands to converge. For each heterostructure, I use approximately 1.5 times the number of electrons in the system as the total number of bands and find that for this number of bands all heterostructure dielectric spectra have converged up to 4.2 eV (the maximum energy value considered in these calculations).

A dense k-point mesh is also required to effectively sample the BZ and accurately determine the optical properties of a system. Here, a $22 \times 22 \times 2$ k-point mesh is used. Both increasing the in-plane k-point mesh to 24×24 and increasing the out of plane k-points from 2 to 4 are found to have no effect on the total energy or dielectric spectrum. An energy cutoff of 350 eV is found to be sufficient for these calculations as well, and no noticeable change is seen when a cutoff of 400 eV is used instead.

3.1 Atomistic Interface Effects

The frequency-dependent absorption coefficient can now be calculated by combining the real and imaginary parts of the dielectric function [30]

$$\alpha(\omega) = \frac{\sqrt{2}\omega}{c}\sqrt{-\epsilon_1(\omega) + \sqrt{\epsilon_1(\omega)^2 + \epsilon_2(\omega)^2}}. \quad (3.3)$$

Once the absorption coefficients and thicknesses for the two supercells are obtained, the macroscopic value is subtracted from the atomistic and divided by two (to account for the fact that there are two interfaces per supercell). The result is a frequency-dependent interface absorbance difference, $\Delta\alpha_i$, expressed as

$$\Delta\alpha_i(\omega) = \frac{A_{\text{Atom}} - A_{\text{Macr}}}{2} = \frac{\alpha_{\text{SC}}(\omega)t_{\text{Atom}} - (\alpha_{\text{NiSi}_2}(\omega)t_{\text{Macr-NiSi}_2} + \alpha_{\text{Si}}(\omega)t_{\text{Macr-Si}})}{2} \quad (3.4)$$

where α_{SC} is the DFT calculated absorption coefficient for the supercell (as defined by Eq. (3.3)), and α_{NiSi_2} and α_{Si} are the DFT absorption coefficients for bulk $NiSi_2$ and Si.

3.1.3 Accuracy of the Generalized Gradient Approximation

It is known that the absorption spectrum for semiconductors is not described adequately by the GGA approximation. More sophisticated calculations such the Heyd, Scuseria, and Ernzerhof range-separated hybrid functional [31, 32], one body Green's function approaches based on the GW approximation [33], and the Bethe-Salpeter Equation [34] are needed for accurate calculations that can be compared with experimental data. However, such calculations are currently impractical for the large supercells that must be used. The calculations here are a proof of concept, designed to demonstrate the potential for extracting and applying interface effects on absorption, as opposed to a rigorous description of $NiSi_2$/Si heterostructure absorption, and are thus left at the GGA level.

In order to validate GGA absorption coefficient used in the calculations, the imaginary part of the dielectric functions, ϵ_2 of the bulk materials is compared to known sources. When calculating $\Delta\alpha_i$, ϵ_2 is the fundamental quantity, as all of the other optical properties are derived from it as shown in Eqs. (3.2)–(3.3). Figure 3.3 shows the calculation for ϵ_2 for $NiSi_2$ compared to reflectivity measurements from Amiotti et al. [35]. The two main peaks (at 2.6 and 4.8 eV) both coincide with the experimental data, indicating that our theoretical calculations for the optical properties of $NiSi_2$ are reasonably accurate.

For Si it is known that GGA underestimates the bandgap, so rather than comparing ϵ_2 for Si against experimental data, it is compared to the first-principles Si dielectric spectrum in the GGA approximation reported by Gajdoš et al. [29] to verify that the calculations are properly converged. The calculated spectrum has peak locations and with magnitudes corresponding to the known values, indicating that it is accurate within the GGA.

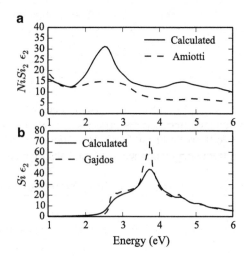

Fig. 3.3 Validation of calculations. (**a**) NiSi$_2$. The main features are the peak at 2.7 eV and the peak at 4.5 eV, which are both present at the same values in experiments. (**b**) Si. Si is calculated within the GGA approximation hence the underestimation of the bandgap. However all other peak values in the Si dielectric spectrum match with the accepted values for GGA-Si

3.1.4 Absorption and Reflection at the Atomic Scale

In order to consider the optical response of multilayer heterostructures, an important limitation of this scheme must be addressed, which is that it cannot be applied to structures where multiple reflections are dominant. In order to meaningfully apply $\Delta\alpha_i$ to a measurable, such as optical transmission, it is required to assume that light interacts with the entire interface region. In the case of multiple reflections, some light interacts more with one side of an interface than the other, and hence $\Delta\alpha_i$ cannot be applied.

The reflectivity of the interface is calculated from the real, η, and imaginary, κ, parts of the refractive index:

$$R_{AB}(\omega) = \frac{\eta_A(\omega) - \eta_B(\omega))^2 + (\kappa_A(\omega) - \kappa_B(\omega))^2}{\eta_A(\omega) + \eta_B(\omega))^2 + (\kappa_A(\omega) + \kappa_B(\omega))^2} \quad (3.5)$$

where η and κ in terms of the dielectric function are

$$\eta(\omega), \kappa(\omega) = \sqrt{\frac{+,-\epsilon_1(\omega) + \sqrt{\epsilon_1(\omega)^2 + \epsilon_2(\omega)^2}}{2}}. \quad (3.6)$$

To demonstrate this limitation, consider a slab of a single material in a vacuum. Once the values for the reflectivity, R, absorption coefficient, α, and length of the slab, l, are known, the fractional transmittivity of the material can be calculated

$$T = (1-R)^2 e^{-\alpha l}(1 + R^2 e^{-2\alpha l} + R^4 e^{-4\alpha l} + \cdots). \quad (3.7)$$

Equation (3.7) assumes that the reflectivity is symmetric on both sides of the slab and that α is homogenous throughout the layer. By breaking down Eq. (3.7) into its

3.1 Atomistic Interface Effects

individual terms, the contribution of the multiple reflections to transmittivity can be determined. The first term in the series is the "first pass" transmittivity, or fraction of light that transmits through the medium without reflection, while each of the subsequent terms in the series refers to an increasing order of multiple reflections. Explicitly, $R^2 e^{-2\alpha l}$ refers to the fraction of light that reflects off the back interface and then off of the front interface before transmitting through the back interface on the "second pass," $R^4 e^{-4\alpha l}$ is the "third pass," and so on and so forth.

Now extend the above equation to a two-layer heterostructure with an interface, such that light transmitting through passes from vacuum (V) into material A, then into material B, and then back into vacuum. The "first pass" transmittivity is defined straightforwardly based off of absorption and the reflections from the surfaces and interfaces

$$T_{\text{fp}} = (1-R_{VA})e^{-\alpha_A t_A}(1-R_{AB})e^{-\alpha_B t_B}(1-R_{BV}) = (1-R_{VA})(1-R_{AB})(1-R_{BV})e^{-(\alpha_A t_A + \alpha_B t_B)}. \tag{3.8}$$

The exponent in Eq. (3.8) is the macroscopic absorbance defined in Eq. (3.4), indicating that it can be corrected using $\Delta\alpha_i$. However, multiple reflections are not so simple. All possible permutations of reflections between the surfaces and interfaces must be considered, and through Taylor expansions, the contributions of all of the reflections can be reduced to the following expression:

$$T = \frac{T_{\text{fp}}}{1 - R_{VA}R_{AB}e^{-2\alpha_A t_A} - R_{AB}R_{BV}e^{-2\alpha_B t_B} - R_{VA}R_{BV}(1-2R_{AB})e^{-2(\alpha_A t_A + \alpha_B t_B)}}. \tag{3.9}$$

The four terms in the denominator of Eq. (3.9) represent, respectively, the "first pass," all internal reflections within layer A, all internal reflections within layer B, and finally all internal reflections between the two edges of the heterostructure. The "first pass" can be corrected using $\Delta\alpha_i$ as previously stated, and the final term also contains the macroscopic absorbance and can be corrected with $\Delta\alpha_i$, but the two middle terms only contain transmissions through one layer, which means $\Delta\alpha_i$ would have to be divided up based off of its contributions from the two sides of the interface. The nature of DFT calculations gives us the response of the entire supercell as a whole, not a position-dependent response, and as a result, such a division is inherently impossible with this method.

However, if the reflectivity at the interface is low, then the "first pass" transmittivity is dominant and reflections play a negligible role. Here, $NiSi_2$/Si interfaces are examined, and Eq. (3.5) can be used to calculate and plot the reflectivity of a $NiSi_2$/Si interface. Figure 3.4 shows the reflectivity throughout the optical regime. While there are some peaks in the reflectivity, at no point does the value go above 10% indicating that multiple reflections should not play a dominant role in superlattices composed of these materials. In addition to $NiSi_2$/Si, many other optical material combinations have similarly low reflectivities, making the method useful in many optical applications [36].

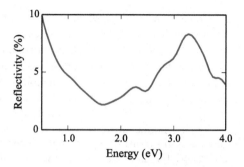

Fig. 3.4 Reflectivity of NiSi$_2$/Si interface. Reflectivity < 10% in optical regime

3.2 Converging the Interface

The next step is to demonstrate that $\Delta\alpha_i$ is truly an independent parameter of a given interface, unrelated to the supercell from which it is calculated. Figure 3.5 shows that as the layer thicknesses in $(NiSi_2)_m/(Si_2)_n$ increases, $\Delta\alpha_i$ converges to a single spectrum, at a layer thickness of 2.57 nm for NiSi$_2$ ($m = 8$) shown in Fig. 3.5a and 1.89 nm for Si ($n = 6$) shown in Fig. 3.5b. The reason for the convergence of $\Delta\alpha_i$ is that, past a critical layer thickness, the interface-induced changes to electrical properties have decayed to their bulk values by center of the slab, as seen in Fig. 3.5c.

Once the electronic properties have returned to those of the bulk, the interface consists of a finite area and $\Delta\alpha_i$ is converged. It is important to note that corrections to account for local-fields and/or electron-hole interactions could impact the $\Delta\alpha_i$ convergence, but since at the convergence thickness the energy bands have reached their bulk values, it is likely that the convergence thicknesses in these advanced schemes are not significantly different. Below the convergence thickness, the interfaces on opposite sides of each layer in the supercell interact with one another and change the absorption. For such "non-interface-converged" heterostructures, the absorbance must be calculated directly using DFT. However, if the layers are thick enough such that interface-interface interactions are negligible, $\Delta\alpha_i$ does not change as layer thicknesses in the heterostructure increase and the calculation of the atomistic absorbance of any "interface-converged" heterostructures can be accessed directly from the converged $\Delta\alpha_i$ added to the macroscopic absorbance (like in Fig. 3.2b).

3.3 Inverted Design Through Interface Concentration

3.3.1 Combining Distinct Interfaces

Ultimately, the goal of the extraction of the interface effects is to apply them to large structures that cannot be calculated directly with DFT. To that end, I demonstrate that one replicates the optical response of a large complex structure with multiple

3.3 Inverted Design Through Interface Concentration

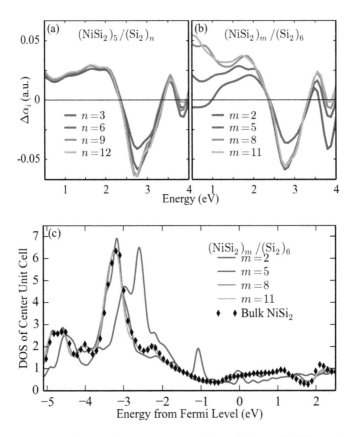

Fig. 3.5 Convergence of $\Delta\alpha_i$. (**a**) Convergence w.r.t. Si layer thickness. Si thickness does not have a significant effect on the absorbance spectrum, and complete convergence occurs by ($n = 6$) or 1.89 nm. (**b**) Convergence w.r.t. NiSi$_2$ layer thickness. For thin layers ($m = 2$), the spectral location of the peak enhancement changes but reaches convergence at ($m = 8$) or 2.57 nm. (**c**) Convergence of the DOS in the center unit cell of the NiSi$_2$ layer. For thin layers, the DOS at the center of the layer differs significantly from the bulk values, but above the convergence thickness, the layer returns to the bulk electronic configuration and the influence of the interface is terminated

interfaces by determining $\Delta\alpha_i$ for each individual interface. I calculate $\Delta\alpha_i$ from several smaller supercells and show that the sum of the individual $\Delta\alpha_i$ is equivalent to $\Delta\alpha_i$ from a large supercell which contains the same interfaces present in the smaller supercells.

The large and complex (but still manageable for DFT) supercell is (NiSi$_2$)$_2$/(Si$_2$)$_3$/(NiSi$_2$)$_5$/(Si$_2$)$_6$. There are four interfaces in this supercell, so I also calculate $\Delta\alpha_i$ for supercells with each of those four interfaces: (NiSi$_2$)$_2$/(Si$_2$)$_3$, (Si$_2$)$_3$/(NiSi$_2$)$_5$, (NiSi$_2$)$_5$/(Si$_2$)$_6$, and (Si$_2$)$_6$/(NiSi$_2$)$_2$. For the small supercells, $\Delta\alpha_i$ has contributions from two identical interfaces and must be reduced by a half with respect to $\Delta\alpha_i$ for the large supercell which contains a single contribution from four distinct interfaces. Figure 3.6 shows $\Delta\alpha_i$ for each of the small supercells,

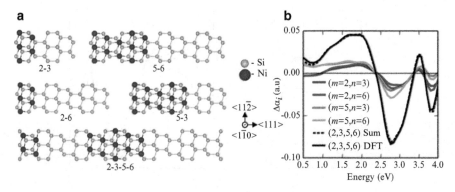

Fig. 3.6 $\Delta\alpha_i$ from individual interfaces vs. $\Delta\alpha_i$ from complex heterostructure. $\Delta\alpha_i$ from individual interfaces vs. $\Delta\alpha_i$ from complex heterostructure. (**a**) Examples of the complex structure (2-3-5-6) and each of the four different $(NiSi_2)_m/(Si_2)_n$ supercells representing the individual interfaces within: (2-6), (5-3), (2-3), and (5-6). (**b**) We calculate $\Delta\alpha_i$ for each supercell shown in (**a**) and then compare the $\Delta\alpha_i$ calculated directly from the (2-3-5-6) supercell to the sum of $\Delta\alpha_i$ from the smaller interfaces to demonstrate that the interface effects of a large system can be determined from the individual interfaces

as well as their cumulative sum, and $\Delta\alpha_i$ from the larger supercell. The excellent agreement between the sum of the smaller supercells and the direct calculation of the larger cell indicates that $\Delta\alpha_i$ accounts for interface absorbance in a consistent, and most importantly, extractable way. Thus, the atomistic absorbance of large complex multilayer heterostructures that would be impossible to calculate directly with DFT can be determined by calculating $\Delta\alpha_i$ for each interface in the structure and adding them linearly.

3.4 Quantitative Applications

3.4.1 Interface Absorption vs. Bulk Absorption

One of the potential applications for this method is to utilize the quantitative nature of $\Delta\alpha_i$ to assess the influence of interfaces on a system. In order to do this, I first define a metric that contains the quantitative value of the interface effects

$$R_{\text{Int:Bulk}}(m, n) = \frac{|\Delta\alpha_i|}{|A_{\text{Macr-}m,n}|}. \tag{3.10}$$

$R_{\text{Int:Bulk}}(m, n)$ is the ratio of $\Delta\alpha_i$ to the macroscopic absorbance. When interface effects are small, $R_{\text{Int:Bulk}}(m, n)$ is small as well. When the interface effects are dominant, $R_{\text{Int:Bulk}}(m, n)$ is high. As a result, $R_{\text{Int:Bulk}}(m, n)$ can be thought of as the relative strength of the interface, in terms of absorption in a $(NiSi_2)_m/(Si_2)_n$ heterostructure.

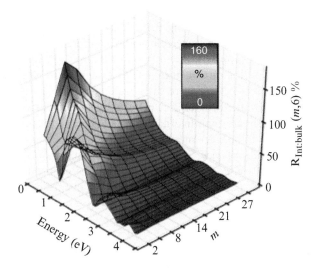

Fig. 3.7 The relative strength of interface absorption. $R_{\text{Int:Bulk}}(m, n)$, the ratio of $\Delta\alpha_i$ and the macroscopic absorbance is determined for a series of $(\text{NiSi}_2)_m/(\text{Si}_2)_6$ heterostructures with varying m. At high thicknesses ($m = 30$), bulk dominates everywhere, except for a small interface effect in the far infrared. At sub-convergence thicknesses ($m = 2$), there is a high interface to bulk ratio, but the $\Delta\alpha_i$ is weaker since it is not converged. The peak relative strength of the interface occurs at the convergence value of $\Delta\alpha_i$, where interface absorption is at its peak, with as little additional bulk as possible. From this result, it can be determined that for larger systems (≥ 10 nm features), interface absorption effects do not play a significant role, but at single digit nanometer thicknesses, interface effects must be accounted for

Figure 3.7 shows a plot of $(\text{NiSi}_2)_2/(\text{Si}_2)_3$, for a range of m values while keeping n at the convergence value of $n = 6$. The plot shows in which thickness regimes interface absorption plays a significant role. The interface effects are primarily at far-IR (low eV), but for sub-convergence thicknesses, $m < 8$, the near-IR/vis (~2 eV) features start to become dominant as well. Recall from Fig. 3.5b that $\Delta\alpha_i$ is weaker at these sub-convergence thicknesses, but because bulk absorbance through such thin layers is weak as well, the interface still plays a dominant role. Interface absorption has its strongest effect at the convergence value for the NiSi_2 layer, $m = 8$, where $\Delta\alpha_i$ has reached full strength, but bulk absorption has not yet begun to dominate. For super-convergence thicknesses, $m > 8$, $\Delta\alpha_i$ stays constant, but bulk absorption increases, reducing the relative strength of the interface effect.

3.4.2 Wavelength Selectivity and Absorption Enhancement

A more advanced application of the method is to use $\Delta\alpha_i$ to rationally design heterostructures with selectively tuned absorption. I use $\Delta\alpha_i$ and the macroscopic absorbance to determine the atomistic absorbance of different supercells, and then normalize them with respect to thickness to reverse engineer the atomistic absorption coefficient of each supercell. In Fig. 3.8a, the atomistic absorption coefficients of $(NiSi_2)_m/(Si_2)_n$ are plotted for supercells with six different configurations, $(m = 2, n = 3)$, $(m = 5, n = 8)$, $(m = 8, n = 12)$, $(m = 10, n = 15)$, $(m = 20, n = 30)$, and $(m = 40, n = 60)$. It is important to note that the ratio of $NiSi_2$ to Si is kept approximately constant (~2:3) in all of these supercells, and the main difference between them is the number of interfaces per unit thickness. The $(m = 2, n = 3)$ and $(m = 5, n = 8)$ cases are "non-interface-converged" structures and require their absorption coefficients to be calculated explicitly. All the others use the converged $\Delta\alpha_i$ and macroscopic absorbance. The strongest absorption enhancement happens in the range of 0.5–2.0 eV, and above these energies the contributions from bulk absorption dominate the spectrum so these regimes are not considered.

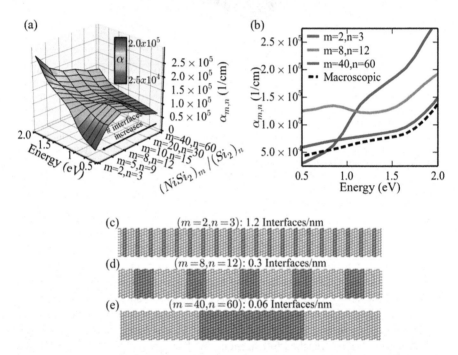

Fig. 3.8 Tunable superlattice absorption. (**a**) The absorption coefficients for $(NiSi_2)_m/(Si_2)_n$ supercells are shown with respect to various values of m and n, keeping the ratio between them constant ~2:3. (**b**) Three distinct spectra appear within the range of structures calculated at different regimes of interface concentration per unit length. (**c**) For sub-convergence structures (i.e. $m = 2, n = 3$), there is a high concentration of interfaces, 1.2 per nm, and as a result, strong interface enhancement in the near-IR/visible. (**d**) For "interface-converged" structures ($m = 8, n = 12$), a new peak emerges in the far-IR, while the strength of the near-IR/visible absorption is diminished due to the lower interface concentration of 0.3 per nm. (**e**) For thick layers $m = 40, n = 60$ the interface concentration is too low to have a significant effect, 0.06/nm, and the atomistic absorption is barely different from the macroscopic absorption

In Fig. 3.8a it can be seen that the supercell absorption spectrum changes significantly as the concentration of interfaces increases. There are essentially three distinct types of spectra in play: Strong enhancement in the visible/near-IR range for thin layers, enhancement in the far-IR range for layers with intermediate thickness, and little to no enhancement for thick layers. Figure 3.8b shows each of these distinct spectra compared against the macroscopic absorption spectrum for a 2:3 ratio of bulk $NiSi_2$:Si.

For structures with thinner layers, such as the $(m = 2, n = 3)$ structure shown in Fig. 3.8c, there is a high concentration of interfaces per unit length, 1.2 per nm. As the layer thicknesses in the heterostructures approach the convergence thicknesses (eight unit cells for $NiSi_2$ and six for Si), the spectral range of the peak enhancement changes. Recall from Fig. 3.5 that at low $NiSi_2$ layer thicknesses, there is little enhancement in the far-IR and suppression at $m = 2$. As a result, the $(m = 2, n = 3)$ structure has its absorption peak at an entirely different spectral location than the $(m = 8, n = 12)$ structure (shown in Fig. 3.8d. The supercells with long periods are dominated by the bulk, such as the $(m = 40, n = 60)$ structure shown in Fig. 3.8e, and the atomistic absorption coefficients are not significantly different from a linear combination of the bulk absorption coefficients. Thus, the method predicts which layer thickness regimes provide absorption enhancement in which spectral ranges and in which regimes the interface absorption effects have a strong effect. By choosing supercell periodicity to take advantage of the changing shape of the interface absorption profile below the convergence value, not only is the absorption in the structure enhanced, the spectral region where the enhancement occurs is selectable.

These calculations provide a new technique in the inverted design of materials with selectable optical properties, due to its ability to generate atomistic accuracy with macroscopic simplicity. It can now be expanded to new systems to help optimize optical detectors and photovoltaics or many other applications as well as simply to assess the impact of interface absorption in complex systems. The quantitative nature of the calculation of $\Delta\alpha_i$ makes it a powerful tool in the design of and study of nanotechnology.

References

1. Hachtel, J.A., Sachan, R., Mishra, R., Pantelides, S.T.: Quantitative firstprinciples theory of interface absorption in multilayer heterostructures. Appl. Phys. Lett. **107**, 091908 (2015)
2. Pasquarello, A., Hybertsen, M.S., Car, R.: Si 2p core-level shifts at the Si(001)-SiO2 interface: a first-principles study. Phys. Rev. Lett. **74**, 1024–1027 (1995)
3. Wang, Z., Wang, J., Sham, T.-K., Yang, S.: Tracking the interface of an individual ZnS/ZnO nano-heterostructure. J. Phys. Chem. C **116**, 10375–10381 (2012). ISSN: 1932-7447
4. Xing, G., et al.: Low-temperature solution-processed wavelength-tunable perovskites for lasing. Nat. Mater. **13**, 476–480 (2014). ISSN: 1476-1122
5. Pavesi, L., Dal Negro, L., Mazzoleni, C., Franzò, G., Priolo, F.: Optical gain in silicon nanocrystals. Nature **408**, 440–444 (2000). ISSN: 0028-0836

6. Rogalski, A.: Quantum well photoconductors in infrared detector technology. J. Appl. Phys. **93**, 4355–4391 (2003). ISSN: 0021-8979, 1089-7550
7. Grätzel, M.: Solar energy conversion by dye-sensitized photovoltaic cells. Inorg. Chem. **44**, 6841–6851 (2005). ISSN: 0020-1669
8. Atwater, H.A., Polman, A.: Plasmonics for improved photovoltaic devices. Nat. Mater. **9**, 205–213 (2010)
9. Sachan, R., et al.: Enhanced absorption in ultrathin Si by NiSi2 nanoparticles. Nanomater. Energy **2**, 11–19 (2013). ISSN: 2045-9831, 2045-984X
10. Shi, N., Ramprasad, R.: Atomic-scale dielectric permittivity profiles in slabs and multilayers. Phys. Rev. B **74**, 045318 (2006)
11. Luo, W., Pennycook, S.J., Pantelides, S.T.: Magnetic "dead" layer at a complex oxide interface. Phys. Rev. Lett. **101**, 247204 (2008)
12. Popescu, V., Zunger, A.: Localized interface states in coherent isovalent semiconductor heterojunctions. Phys. Rev. B **84**, 125315 (2011)
13. Pinchuk, A., Kreibig, U., Hilger, A.: Optical properties of metallic nanoparticles: influence of interface effects and interband transitions. Surf. Sci. **557**, 269–280 (2004). ISSN: 0039-6028
14. Oubre, C., Nordlander, P.: Optical properties of metallodielectric nanostructures calculated using the finite difference time domain method. J. Phys. Chem. B **108**, 17740–17747 (2004). ISSN: 1520-6106
15. Catchpole, K.R., Polman, A.: Design principles for particle plasmon enhanced solar cells. Appl. Phys. Lett. **93**, 191113–191113-3 (2008). ISSN: 00036951
16. Carrier, P., Lewis, L.J., Dharma-wardana, M.W.C.: Optical properties of structurally relaxed Si/SiO_{2} superlattices: the role of bonding at interfaces. Phys. Rev. B **65**, 165339 (2002)
17. Giustino, F., Umari, P., Pasquarello, A.: Dielectric discontinuity at interfaces in the atomic-scale limit: permittivity of ultrathin oxide films on silicon. Phys. Rev. Lett. **91**, 267601 (2003)
18. Luppi, M., Ossicini, S.: Ab initio study on oxidized silicon clusters and silicon nanocrystals embedded in SiO_{2}: beyond the quantum confinement effect. Phys. Rev. B **71**, 035340 (2005)
19. Giustino, F., Pasquarello, A.: Infrared spectra at surfaces and interfaces from first principles: evolution of the spectra across the Si(100)-SiO_{2} interface. Phys. Rev. Lett. **95**, 187402 (2005)
20. Klipstein, P.C., et al.: A k . p model of InAs/GaSb type II superlattice infrared detectors. In: Infrared Physics & Technology. Proceedings of the International Conference on Quantum Structure Infrared Photodetector (QSIP) 2012, vol. 59, pp. 53–59 (2013). ISSN: 1350-4495
21. Brezesinski, T., Wang, J., Polleux, J., Dunn, B., Tolbert, S.H.: Templated nanocrystal-based porous TiO2 films for next-generation electrochemical capacitors. J. Am. Chem. Soc. **131**, 1802–1809 (2009). ISSN: 0002-7863
22. Milliron, D.J., Buonsanti, R., Llordes, A., Helms, B.A.: Constructing functional mesostructured materials from colloidal nanocrystal building blocks. Acc. Chem. Res. **47**, 236–246 (2014). ISSN: 0001-4842
23. Barg, S., et al.: Mesoscale assembly of chemically modified graphene into complex cellular networks. Nat. Commun. **5** (2014/2015). https://doi.org/10.1038/ncomms5328. http://www.nature.com.proxy.library.vanderbilt.edu/ncomms/2014/140707/ncomms5328/full/ncomms5328.html
24. Llordés, A., Garcia, G., Gazquez, J., Milliron, D.J.: Tunable near-infrared and visible-light transmittance in nanocrystal-in-glass composites. Nature **500**, 323–326 (2013). ISSN: 0028-0836
25. Kresse, G., Hafner, J.: Ab initio molecular dynamics for liquid metals. Phys. Rev. B **47**, 558–561 (1993)
26. Kresse, G., Hafner, J.: Ab initio molecular-dynamics simulation of the liquidmetal-amorphous-semiconductor transition in germanium. Phys. Rev. B **49**, 14251–14269 (1994)
27. Blochl, P.E.: Projector augmented-wave method. Phys. Rev. B **50**, 17953–17979 (1994). ISSN: 1098-0121

References

28. Perdew, J.P., Burke, K., Ernzerhof, M.: Generalized gradient approximation made simple. Phys. Rev. Lett. **77**, 3865–3868 (1996)
29. Gajdoš, M., Hummer, K., Kresse, G., Furthmüller, J., Bechstedt, F.: Linear optical properties in the projector-augmented wave methodology. Phys. Rev. B **73**, 045112 (2006)
30. Fox, M.: Optical Properties of Solids. Oxford University Press, Oxford (2010)
31. Heyd, J., Scuseria, G.E., Ernzerhof, M.: Hybrid functionals based on a screened Coulomb potential. J. Chem. Phys. **118**, 8207–8215 (2003). ISSN: 0021-9606, 1089-7690
32. Heyd, J., Scuseria, G.E., Ernzerhof, M.: Erratum: "Hybrid functionals based on a screened Coulomb potential" [J. Chem. Phys. **118**, 8207 (2003)]. J. Chem. Phys. **124**, 219906 (2006). ISSN: 0021-9606, 1089-7690
33. Hedin, L.: Something to do with GW. Phys. Rev. **139**, A796 (1965)
34. Rohlfing, M., Louie, S.G.: Electron-hole excitations and optical spectra from first principles. Phys. Rev. B **62**, 4927–4944 (2000)
35. Amiotti, M., Borghesi, A., Guizzetti, G., Nava, F.: Optical properties of polycrystalline nickel silicides. Phys. Rev. B **42**, 8939–8946 (1990)
36. Zhu, L., Luo, J.K., Shao, G., Milne, W.I.: On optical reflection at heterojunction interface of thin film solar cells. Solar Energy Mater. Solar Cells **111**, 141–145 (2013). ISSN: 0927-0248

Chapter 4
Advanced Electron Microscopy for Complex Nanotechnology

From this point forward, the focus of this thesis will be on electron microscopy. In this chapter, I examine complex nanostructures with applications in nanotechnology that are highly dependent on morphological, structural, compositional, and optical effects. The electron microscope is the ideal tool for this kind of analysis, and I show here a wide range of different STEM techniques that can be used to characterize complex nanotechnology with nanoscale precision.

4.1 Ge-Based FET Devices

Germanium-based devices have been of significant interest in the field of solid state devices since the field was founded; in fact the first generation of transistors that won Nobel Prizes for Bardeen, Shockley, Brittain, and Killby were all made of Ge not Si [1–4]. However, Ge did not thrive in the age of MOSFETs as the native oxides to Ge had significantly inferior electrical performance compared with Si's native oxides [5, 6]. Recently, however, Ge has returned to the forefront of CMOS technology due to the advent of high-k dielectrics and the utilization of multilayer gate stacks [7–10].

For the first section of this chapter, STEM is used to investigate a number of different Ge-based FET designs, with the goal of understanding the structural and compositional properties of the devices. All devices in this section are fabricated at Interuniversity Microelectronics Center (IMEC) in Leuven, Belgium, and prepared by me for STEM analysis using the Zeiss Auriga dual-beam FIB/SEM at the University of Tennessee at Knoxville. Details of the dual-beam extraction process are outlined in Appendix C.1.

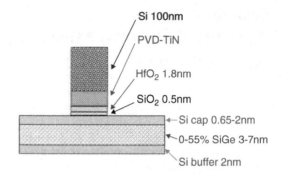

Fig. 4.1 Si-capped $Si_{0.55}Ge_{0.45}$ pMOSFET schematic. Cross-sectional view of the gate stack in the device with the predicted layer thicknesses

4.1.1 Negative-Bias Temperature Instability in Flat Si-Capped pMOSFETs

Negative-bias temperature (NBT) instabilities are one of the primary reliability concerns in all highly scaled Si-based pMOSFETs [11–13], because the interface-trap charges generated during NBT stress detrimentally affect device performance [14, 15]. This section deals specifically with the charge-trapping properties of $Si_{0.55}Ge_{0.45}$ pMOSFETs fabricated on an n-type Si wafer, with a Si-capping layer to separate the channel from gate oxides (schematic shown in Fig. 4.1).

Electrical characterization of the samples was performed by Dr. Guoxing Duan and presented in [16] and is compared against measurements from Si pMOS FinFETs presented by Manmouni et al. in [17] as a reference.

The device performance is measured by transconductance, which is the ratio of the output current of the device to the input voltage, $g_m = \frac{I_{out}}{V_{in}}$. In Fig. 4.2a the transconductance of the sample pre- and post-NBT stress is shown, and the effects of NBT instabilities on the device can be observed. Both the threshold voltage and the peak transconductance are affected by the stress, and the interface-trap charge density can play a role in the degradation of the device. In Fig. 4.2b the effective interface-trap generation, ΔN_{IT}, taken as a function of temperature for the SiGe devices stressed at -11.1 MV/cm is compared to ΔN_{IT} from the reference Si FinFETs stressed at -10.3 MV/cm [16, 17]. Here it can be seen that the interface-trap charge densities are higher in the SiGe devices to the Si reference, along with having a lower activation energy (0.14 eV-SiGe, 0.25 eV-Si).

Interface-trap generation similar to the values in these SiGe MOSFETs has been observed in Si MOS transistors under NBT stress. It was seen to be a result of the hydrogen atoms, attached to dopants that drift to the interface, depassivating the interface Si-H bonds by forming H_2 molecules [11, 18]. However, assessing the validity of such a claim requires high-resolution structural and compositional studies, so a cross-sectional STEM analysis is applied to the devices.

STEM provides an atomistic picture of the Si-capped SiGe MOS structure and the distribution of the atoms within. Figure 4.3a and b show HAADF and ABF images of the gate stack showing the distinct layers in the device. Two of the

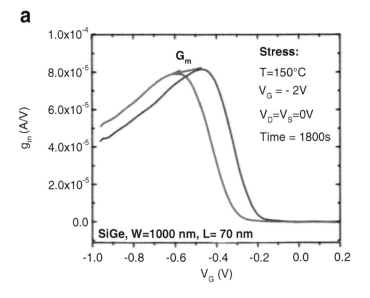

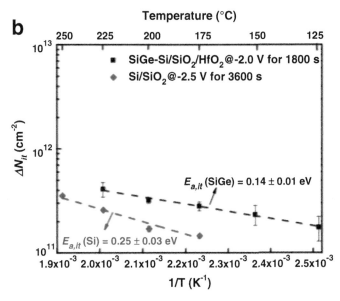

Fig. 4.2 Negative-bias temperature instabilities in SiGe pMOSFETs. (**a**) Transconductance measurements as a function of gate voltage (V_G) at room temperature before and after 30 min of -2 V stress on the gate at a temperature of 150 °C, showing reduced peak transconductance and shifted threshold voltage. (©2015 IEEE. Reprinted with permission from [16].) (**b**) Arrhenius plots of the effective interface-trap generation in the SiGe devices (black) compared to the reference Si FinFETs (blue). The SiGe device suffers both from lower activation energies for interface traps as well as a higher interface-trap charge density

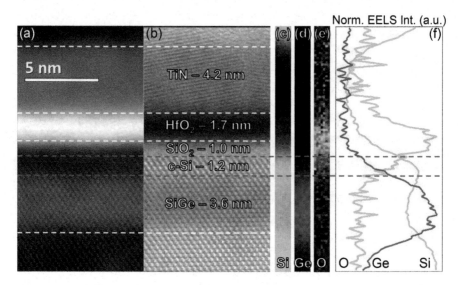

Fig. 4.3 STEM analysis of SiGe MOS gate stack. (**a**) HAADF (©2015 IEEE) and (**b**) ABF cross-sectional images of the SiGe pMOSFET with HfO$_2$/SiO$_2$ gate dielectrics. The 1.4 nm Si cap is partially oxidized, yielding a ~1 nm SiO$_2$ interfacial layer and a ~1 nm thick Si-capping layer. (**c**)–(**e**) Two-dimensional STEM-EELS chemical maps of the device for the elements in the dielectric/capping/channel layers. (**f**) The EELS intensities in (**c**)–(**e**) are horizontally binned, normalized, and plotted together. From the EELS data, it can be seen that there is significant Ge diffusion through the Si-capping layer (dashed red lines) all the way up to the oxide layer to allow for the depassivation of Si-H bonds at this interface. ©2015 IEEE

common concerns for devices with the SiGe/Si/SiO$_2$/high-k dielectric structure are the complete oxidization of the Si-capping layer and the strain of the Ge in the Si lattice resulting in dislocation defects in the crystalline capping layer. However, the atomic-resolution images preclude both possibilities, as the images show the amorphous SiO$_2$ layer (~1 nm) and the unconsumed, crystalline Si layer (~1 nm) distinctly, as well as showing that the crystallinity in the Si-capping layer is uniform across the entire channel (indicating the Ge-alloyed layer has not exceeded the critical relaxation thickness) [19].

EEL-SI are taken of the gate stack, and the two-dimensional maps are plotted in Fig. 4.3c through 4.3e, Si K-edge at 1839 eV, the Ge L-Edge at 1217 eV, and the O K-edge 532 eV. From these SI, it can be seen that the Ge signal is not restricted to the channel and significant Ge diffusion can be observed in the Si-capping layer. By horizontally binning the SIs, normalizing the intensities, and plotting them jointly, it can be seen that the Ge diffusion extends up to the oxide layer. Density functional theory calculations have shown that the presence of Ge reduces the reaction energy for hydrogen desorption causing the interface-trap charge buildup [20].

4.1 Ge-Based FET Devices

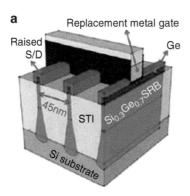
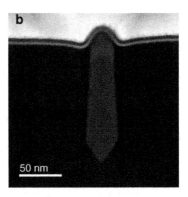

Fig. 4.4 Structure of Ge *p*MOS FinFET reproduced with permission from [25]. (**a**) A schematic showing the three-dimensional geometry of the Ge FinFETs. (**b**) Cross-sectional HAADF image of the resulting Fin

4.1.2 Structural and Compositional Study of Ge pMOS FinFETs

Developments in solid-state fabrication have allowed for the production of novel transistor geometries (such as gate-all-around FETs and FinFETs) to enhance device performance [21, 22]. Since the mobility enhancement from the field effect is limited to the area very close to the gate, the concept of having multiple gate regions enhances the effectiveness of devices [23]. However, fabrication of three-dimensional geometries with nanometer precision has presented a whole new range of technical challenges, making high-resolution analytical techniques like STEM highly valuable [24].

Figure 4.4a shows a schematic of a prototype FinFET being developed by IMEC. Unlike the flat *p*MOSFETs in Sect. 4.1.1, these samples were sent without a known issue to be resolved, and the electron microscopy was intended as an exploratory study of compositional and structural effects. Cross sections of the devices were extracted and analyzed in STEM, and a HAADF image of one of the fins is shown in Fig. 4.4b. The high-resolution STEM image clearly shows many different layers in the gate stack, but with many different elements present, the Z-contrast does not provide the desired level of structural information.

To resolve the layers of the gate structure with compositional (as well as structural) information and nanoscale precision, EEL spectrum imaging is performed. Figure 4.5 shows the various SI for the elements present in the FinFET. The EELS edges used are the (a) Si K-edge at 1839 eV, (b) the Ge L-edge at 1217 eV, (c) the O K-edge 532 eV, (d) the Hf M-edge at 1662 eV, (e) the Ti L-edge and the N K-edge integrated together at 456 eV and 401 eV, respectively, and (f) the W M-edge at 1809 eV. The composite image of all six spectrum images can be seen in Fig. 4.5g showing six distinct regions: the Ge channel, a Si-rich outer layer to the channel, the SiO_2 layer between the gate and the channel, the HfO_2 gate oxide, the TiN gate, and the W via. The scale bar shows that layer thicknesses of order ~1 nm can be resolved and give a high-precision map of the gate-stack region.

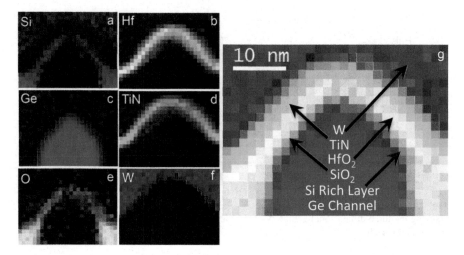

Fig. 4.5 Elemental composition of Ge pMOS FinFET. (**a**)–(**f**) SI maps of the various elements present in the FinFET. (**a**) Si, (**b**) Ge, (**c**) O, (**d**) Hf, (**e**) TiN, (**f**) W. (**g**) Shows a composite image of the six SI in (**a**)–(**f**). From this all the layers of the gate stack are revealed with nanometer precision. ©2017 IEEE. Reprinted with permission from [25]

It is also worth noting that in Fig. 4.5a there is some Si intensity in the W via region. This is not a genuine signal but an artifact. Since the W M-edge is at 1809 eV and has a gradual onset, the W background cannot be adequately subtracted for the sharp onset for the Si K-edge. As a result, the background subtraction used to pick out the Si signal at 1839 eV detects some intensity in the presence of W. However, no Si is present in the W via.

Beyond the composition of the FinFETs, the crystallinity of the structures is highly important. The Si buffer layer prevents the formation of native oxides that can cause parasitic effects, but the defects and inhomogeneities in the layer can also result in the reduction of device efficiency. To examine the crystallinity of the device, a type of Fourier analysis may be employed.

A fast Fourier transform (FFT) is an algorithm for calculating the discrete Fourier transform of a function [26]. The Fourier transform is fully reversible, meaning the inverse FFT (iFFT) of an FFT returns the original image. The reversibility can be exploited by masking the FFT before performing the iFFT allowing for different Fourier components of the original image to be enhanced or filtered out.

Figure 4.6 shows how an iFFT can be used to map the real-space location of a specific periodicity. Figure 4.6a shows a square image with two distinct periodicities; the FFT of the image is shown in Fig. 4.6b. Four bright points in the FFT can be observed, two corresponding to periodicity in the x-dimension and two for the y-dimension. A magnified view of the FFT in Fig. 4.6b is shown in Fig. 4.6c; here it can be seen that the FFT is not just four spots but a wide range of Fourier components with varying intensities. The relative intensity of the surrounding points is altered by the location of the periodicity in the real-space image and allows the

4.1 Ge-Based FET Devices

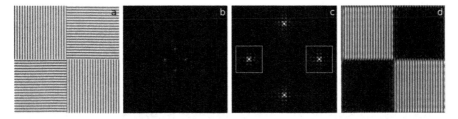

Fig. 4.6 Fourier analysis for mapping periodicities in images. (**a**) A real-space image, showing different periodicities in different spatial regions. (**b**) A fast Fourier transform (FFT) of the image in (**a**). (**c**) To map a periodicity, the Fourier components outside the desired periodic features (outside the blue box) are masked out of the FFT. (**d**) By performing an inverse FFT on the masked FFT, a real-space map of the periodic feature is generated

location of the different periodicities to be detected. In order to map the physical location of the x-dimension periodicity, a mask is applied to the FFT to remove all Fourier components outside the masked region (blue box in Fig. 4.6c). Figure 4.6d shows the iFFT of the masked FFT in Fig. 4.6c and only displays the regions of the image in Fig. 4.6a that possessed the x-dimension periodicity.

By applying this same method to images of the FinFETs, maps of the crystalline features in the sample can be generated. Here, BF images are used as opposed to HAADF images, because BF images have a higher signal-to-noise ratio; while phase contrast makes direct interpretation of the images difficult, the periodic characteristics of the image is maintained.

Figures 4.7a and b show a high-resolution BF image of one of the FinFETs and the corresponding FFT which shows some very strong periodic components. First, the main periodic feature is analyzed through the same iFFT process shown in Fig. 4.6. The result is shown in Fig. 4.7c and d, where Fig. 4.7c is the BF image overlaid with the iFFT map and Fig. 4.7d is the FFT of the BF image and the mask used for forming the iFFT map. The main crystalline pattern is the ⟨110⟩ axis from the Ge channel, which can be confirmed from the iFFT as this pattern is entirely localized to Ge channel region. It is important to note that, unlike the flat devices in Sect. 4.1.1, the Si buffer layer is neither crystalline nor epitaxial with the channel. The crystallinity of the buffer layer is important because defects near the channel can scatter minority carriers, increasing their mean free path and reducing transport efficiency.

Furthermore, the buffer can even be shown to be amorphous. In Fig. 4.7e and f, the iFFT of an annular mask is shown. Annular masks are chosen because amorphous materials still retain some short-range ordering; the result in the FFT is a diffuse ring with roughly same radial magnitude as the lattice spacing of the crystalline version of the material. In Fig. 4.7f, this diffuse amorphous ring is selected with an annular mask, and the Fourier components of the crystalline structures are blocked out of the resulting mask (view of the final mask shown in the inset). The result of the iFFT in Fig. 4.7e shows that the amorphous ring comes from both the surrounding isolation oxide, as well as the buffer between the gate and the channel, indicating that there is no crystalline presence there.

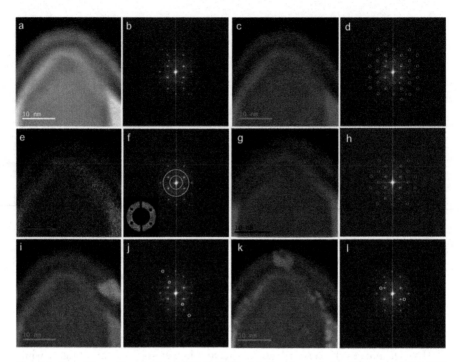

Fig. 4.7 Crystalline effects in Ge *p*MOS FinFETs. (**a**) Atomic-resolution BF image of FinFET. (**b**) FFT of (**a**). (**c**) iFFT map of predominant crystalline feature in (**a**) and (**b**), originating from Ge channel. (**d**) FFT from (**b**) with masked regions used for iFFT in (**c**) marked. (**e**) iFFT and (**f**) FFT for amorphous material in the sample. The diffuse amorphous ring in the FFT pattern is selected with an annular mask, and then individual strong Fourier components are additionally masked; the masked FFT used to calculate (**e**) is shown in the inset of (**f**). (**g**) iFFT and (**h**) FFT for a different fin then the fin used for (**a**)–(**e**). Here, the crystallinity in the channel shows inhomogeneities, but a large portion of the buffer is epitaxial to the channel, unlike the buffer for the fin in (**a**)–(**e**). Returning to the original fin from (**a**)–(**e**), it can be seen that there are Fourier components that are not part of the channel. (**i**) and (**k**) iFFTs and (**j**) and (**l**) FFTs of these spots show they are crystalline grains in the TiN gate layer

However, it is important to remember that the FinFETs are three-dimensional and extend for lengths on the micron scale in the z direction (relative to the STEM images). Since the samples are extremely thin cross-sections, ~70 nm, it is possible that different crystalline features are present at different points in the device. To account for this, different FinFETs were analyzed, and the iFFT of the main Ge channel crystallinity (analog to Fig. 4.7c and d) is shown in Fig. 4.7g and h for a different fin. The crystallinity of this fin, Fig. 4.7g, on the left side of the channel is not as clear as the fin in Fig. 4.7c. Additionally, on the opposite side of the fin, the ⟨110⟩ pattern can be seen to extend up all the way through the buffer up to the gate. The comparison of Fig. 4.7c and g shows that the buffer of the FinFETs is at least partially crystalline, but significant inhomogeneities exist both in the buffer and the channel that likely impact device performance.

On a final note, there are many Fourier peaks visible in Fig. 4.7b that are not associated with the ⟨110⟩ crystallinity of the Ge channel. The iFFTs of these patterns are shown in Fig. 4.7i–l and originate in the TiN gate region. The crystallinity of the metal gate is important to assess because it can significantly affect the surface potential experienced by the minority carriers flowing from the source to the drain, which can also reduce transport efficiency [27, 28].

4.2 Magnetic and Plasmonic Nanocomposites

Beyond solid-state electronics devices, STEM is ideal for studying multicomponent colloidal systems. Nanocomposites with multiple components and multiple properties can be synthesized quite straightforwardly using colloidal techniques with selectable, complex morphologies for a wide range of biomedical applications. Superparamagnetic iron oxide nanoparticles (SPIONs) are biocompatible and extensively used in nanomedicine, and by coupling them with plasmonic gold nanoparticles (Au NPs), also common to biomedicine, a wide range of applications can be accommodated, including imaging, sensing, drug and gene delivery, as well as in photothermal therapies [29–32]. The figures and analysis in this section have been reproduced and adapted from Refs. [33] and [34] by permission of The Royal Society of Chemistry.

The particles are synthesized by Dr. Siming Yu at Intitut de Ciéncia de Materials de Barcelona in the group of Professor Anna Roig. To synthesize these composites, an iron acetylacetonate precursor is added to a solution of polyvinylpyrrolidone (PVP). After thorough dispersion through sonication, the Fe/PVP solution undergoes heating in a microwave reactor to nucleate the SPIONs. By changing the parameters of the microwave heating, average size of the SPIONs can be adjusted between 5 and 10 nm. The PVP/PVP-SPION solution is then mixed with hydrogen tetrachloroaurate, HAuCl4, resulting in the formation of hybrid SPION/Au structures [33]. The morphology and size of the structure can be highly controlled using the molar ration of the PVP to HAuCl4, as well as the duration and temperature of the microwave heating. Dr. Yu observed that the gold morphologies in the samples studied consisted of 60% planar nanotriangles (NT) and 15% of planar nanohexagons (NH), while the remaining nanoparticles were a mixture of platonic structures (23%) and smaller rounded particles (2%), with a 10 wt% magnetic fraction (Fig. 4.8).

4.2.1 Composition of Nanocomposite Components

To investigate the composition of the nanocomposites, EEL spectrum imaging was performed and then analyzed using multiple linear least square (MLLS) fitting. Figure 4.9 displays the composition map of an Au NT by MLLS-EELS mapping.

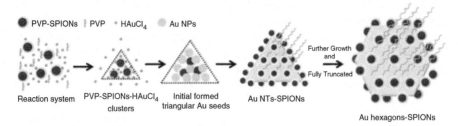

Fig. 4.8 Au/SPION synthesis schematic. HAuCl4 is added to a solution of PVP/SPIONs, the SPIONs stabilize the Au in the solution and cause the nucleation of planar nanostructures such as nanotriangles (NT) and nanohexagons (NH)

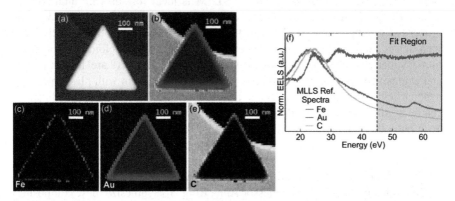

Fig. 4.9 Composition of nanocomposites. (**a**) High-angle annular dark-field (HAADF) image of an Au NT-SPION composite. (**b**) The elemental composite of the Au NT-SPIONs composed of the EEL spectrum image maps of Fe, Au, and C (shown in (**c**)–(**e**), respectively) determined through multiple linear least squares (MLLS) fitting. (**f**) The MLLS reference spectra used for the fitting. Fit region is 45 eV through 65 eV to avoid the similar bulk plasmons of Fe and C

Figure 4.9a shows a HAADF image of the nanoparticle, and Fig. 4.9b shows the elemental composition, where red is iron, blue is gold, and yellow is carbon. Figure 4.9c–e shows each of the individual compositional maps, respectively. Figure 4.9f contains the reference spectra used for MLLS fitting. The fit region is from 45 to 65 eV due to similar positions of the bulk plasmons for iron and carbon in the 15–40 eV band. Due to the presence of the prominent Fe M-edge at 54 eV and the significantly different low-loss character of gold, the 45–65 eV range presents three completely different spectra that are suitable for an MLLS fit.

It is important to note that both the gold and iron maps show high intensities along the edges and weak/no intensity in the bulk. This is due to the thickness of the gold in the bulk of the nanoparticle, which scatters a large portion of the beam to away from the EELS detector. As a result, in thick samples with strongly scattering materials such as gold, intensity can be traced to a reduced number of total electrons in the EEL spectrum, as opposed to a drop in intensity of an individual EEL feature. While the Au NTs and NHs are planar, they possess a three-dimensional character

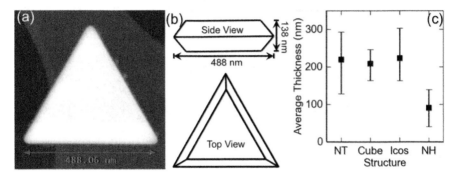

Fig. 4.10 Thickness in the nanostructures. (**a**) HAADF image of NT showing side length of 488 nm. (**b**) Side and top view of NT with EELS thickness measurements and side length shown. (**c**) EELS thickness measurements of the different shapes resulting from the synthesis process

and can have significant thickness variations. EELS low-loss spectra can be used to determine the thickness in terms of the inelastic mean free scattering length (λ) [35]. In Figure 4.10a the NT from Fig. 4.9 in the main text is shown. The side length is determined to be 488 nm from the calibrated image, and the thickness from the low-loss measurements is determined to be 138 nm. A two-scale schematic of the NT is shown in Fig. 4.10b to demonstrate the approximate three-dimensional morphology, where the angles of the facets on the sides are determined through the assumption that the top planar facet is the $\langle 111 \rangle$ and that the facets along the edge are alternating between $\langle 111 \rangle$ and $\langle 100 \rangle$ facets. Figure 4.10c shows low-loss EELS thickness measurements from a large range of NTs and NHs along with more common three-dimensional shapes resulting from the synthesis process: icosahedra and cubes. It can be seen that the NTs are generally thicker than the NHs and approximately as thick as the three-dimensional objects. The planar structures, such as the NHs and NTs, have a three-dimensional character and can become quite thick; EELS log-ratio thickness measurements show that the NTs are 138 nm thick. Such thicknesses are sufficient to result in reduced signal in the nanoparticle across the entire EEL spectrum.

Gold nanocrystals grown through polyol synthesis are expected to be single crystals. In the thinner, hexagonal structures, the STEM can be used with a defocused "Ronchigram" (the zeroth order disk of the convergent beam diffraction (CBED) pattern) to assess crystallinity. In the Ronchigram, diffraction in the sample results in a Kikuchi pattern [36]. Figure 4.11a shows a HAADF image of an Au NH-SPION nanostructure. The sample is tilted so that the $\langle 111 \rangle$ zone axis is parallel to the beam, and the Kikuchi pattern for the $\langle 111 \rangle$ zone axis is shown uninterrupted across the entire NH in Fig. 4.11b, indicating that the entire structure is a single crystal. For the thicker structures where the Kikuchi pattern is not easily observed, crystallinity can be determined through CBED analysis.

Figure 4.11c shows a HAADF image of a thicker structure where no strong Kikuchi pattern is observed. Figure 4.11d and e shows CBED patterns from the

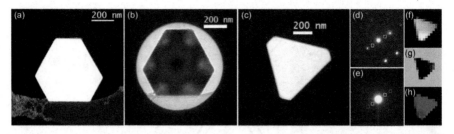

Fig. 4.11 Crystallinity of gold structures. (**a**) HAADF image of an Au NH-SPION structure. (**b**) STEM Ronchigram of the NH aligned such with the ⟨111⟩ crystal axis parallel to the beam. The ⟨111⟩ Kikuchi diffraction pattern can be observed uninterrupted across the entire structure, indicating that the NH is a single crystal. (**c**) HAADF image of an Au NT, too thick for observing the Kikuchi pattern. (**d**) and (**e**) CBED pattern from the Au NT (**d**) and the support C film (**e**) for the Au NT of (**c**). (**f**)–(**h**) 16 × 16 maps of the nanotriangle are made using the HAADF intensity (**f**), the intensity at the amorphous C diffraction ring (**g**) (green squares in (**d**) and (**e**)), and (**h**) the Au ⟨111⟩ diffraction spots (red squares). The entire NT shows the same ⟨111⟩ diffraction pattern

gold bulk and the thin carbon supporting film. A 16 × 16 region including the NT is scanned, and the HAADF intensity along with CBED patterns is collected at each position. Figure 4.11f–h shows the resulting maps, where Fig. 4.11f shows the HAADF map, Fig. 4.11g shows the map from the amorphous carbon diffraction ring seen in Fig. 4.11e (green squares), and Fig. 4.11h shows the most intense diffraction spots from the gold ⟨111⟩ pattern (red squares). The maps show that the same ⟨111⟩ CBED pattern is present at all points in the structure and no shift in the diffraction pattern is observed, indicating that the entire NT is a single crystal.

Finally, EELS fine structure and quantitative analysis are used to determine the phase of the iron oxide present in the SPIONs. Figure 4.12a and b shows the O K- and Fe L-edges taken from three different samples: a reference sample of Fe_2O_3 (with 3+ valence), a reference sample of FeO (with 2+ valence), and finally the SPIONs. In Fig. 4.12a the O K-edge from the three samples is shown. The pre-peak at 529 eV shows a clear distinction between Fe_2O_3 and FeO and, more importantly, that the pre-peak in the SPIONs is almost an exact match of the Fe_2O_3.

Additionally, in the Fe L-edge EELS shown in Fig. 4.12b, it can be seen that the L_3 peak energy of Fe_2O_3 and FeO differs significantly (2+ at 706 eV, 3+ at 708 eV) and that the SPION peak clearly aligns with the Fe_2O_3 peak at 708 eV. Moreover, it is also observed that the FeO L_2-edge has only a single peak at 718 eV while both the SPIONs and the Fe_2O_3 have two peaks at 719 and 721 eV. There are some dissimilarities between the Fe_2O_3 reference sample and the SPIONs, namely, a stronger pre-peak on the L_3 edge and a larger area beneath the L_2 edge. The likely reason for these differences is sample damage from the electron beam.

However, many other iron oxide phases besides Fe_2O_3 and FeO exist and in the absence of reference samples for each of those phases, quantitative EELS is used to further investigate the composition of the SPIONs. The relative concentration is determined by removing the power-law EELS background and then integrating the intensity of the Fe and O EELS cross sections [37]. The near-edge fine structure varies significantly across different compounds, so these spectral regions are excluded from the integration.

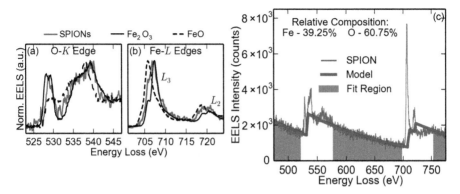

Fig. 4.12 Phase of iron oxide in SPIONs. (**a**) O K-edge EELS. The two reference samples (Fe_2O_3-solid, FeO-dashed) show two distinct behaviors in the relative intensity of the pre-peak at 529 eV, the SPIONs (red) match the Fe_2O_3. (**b**) Fe-L edge EELS. A 2 eV shift (708 eV-Fe_2O_3, 706 eV-FeO) is observed in the peak intensity of the L_3 edge, and the SPIONs share the Fe_2O_3 at 708 eV. (**c**) Relative composition of O and Fe. By fitting X-ray photo-absorption (XRPA) cross sections to the tails of the O and Fe edges, the intensities of the two edges can be directly compared and the relative composition calculated. The result is a 1.548 ratio between O and Fe, further confirming the SPIONs are in the Fe_2O_3 phase

X-ray photo-absorption (XRPA) cross sections are known to be excellent analogues of the corresponding EELS cross sections, and the intensity of the EELS signal is determined by fitting these XRPA cross sections to the power-law tails of the EELS edges [38]. A quantitative model of the EELS signal is formed and plotted against the EELS signal in Fig. 4.12c, where it can be seen that the model matches well the power-law tails of the edges and ignores the variations due to fine structure. From the quantitative model, a relative composition of 39.25% Fe and 60.75% O is determined, which corresponds to the composition of Fe_2O_3 and is in agreement with the analysis in Fig. 4.12a and b. These results indicate that the Fe retains its Fe(III) character and does not reduce to Fe(II). The EELS quantification in this section was performed with the Quantifit code available at http://web.utk.edu/~gduscher/Quantifit/.

4.2.2 Bonding of SPIONs to the Au Nanostructures

Of additional interest, the structural study of the nanocomposites is the bonding of the SPIONs to the surface of the Au substructure. Even though thickness effects make EELS studies over the bulk of the nanoparticle difficult, the fact that faceted edges of the NT taper to a point means that the elemental composition of the Au NT-SPIONs can be observed and analyzed at points of reduced gold thickness. Figure 4.13a shows the same NT from Figs. 4.9 and 4.10 in (a) with a zoomed-in view of the faceted edge in (b). Three regions are indicated: the SPION, the faceted

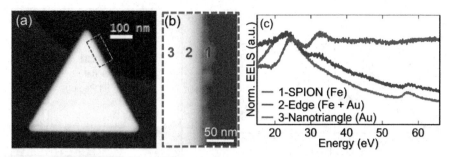

Fig. 4.13 SPION coverage of Au structures. (**a**) HAADF image of the Au NT from Figs. 4.9 and 4.10 with a region around the edge outlined. (**b**) Close-up of the outlined region in (**a**), showing the three regions: the Fe SPION, the faceted edge of the structure, and the thick bulk of the NT. (**c**) Spectra from each of the three regions highlighted in (**b**), showing that the signal from the faceted edge is a combination of Fe and Au and that the coverage of the nanoparticles is relatively uniform across the structure surface

edge, and the bulk of the NT. Figure 4.13c shows low-loss EEL spectra from each of those three regions. Points 1 and 3 each show the characteristic low-loss spectra of iron oxide and gold, respectively; however, at Point 2 the low-loss spectrum appears to be a linear combination of the iron and gold spectra. This indicates that even where the iron EELS signal cannot be detected due to strong gold signal, there are SPIONs present on all facets of the NT, not just on the edges where they are most clearly observable through STEM.

While present on all facets of the nanostructures, it is expected that during crystal growth, free PVP and PVP-SPIONs selectively adsorb on the ⟨111⟩ facets rather than on the ⟨100⟩ facets [39]. In the NH, the crystal facets enclosing the hexagon are the ⟨111⟩ and ⟨100⟩ families in an alternating sequence. Figure 4.14 depicts how the preferential bonding of PVP-SPIONs to the ⟨111⟩ facets is empirically determined. The system is imaged in two tilt orientations, termed max and min tilt. In each orientation the sample is tilted to one extreme, which causes one specific facet of the NH to be closer to parallel to the beam, allowing that facet to be imaged effectively. Since the facets are alternating, if the ⟨100⟩ is aligned with the beam at max tilt (as seen in Fig. 4.14a), then the ⟨111⟩ is aligned with the beam at min tilt (Fig. 4.14b). The tilt is performed such that two facets can be seen at both max and min tilt, and due to the alternating pattern of the ⟨100⟩ and ⟨111⟩ facets, this results in the imaging of two different ⟨100⟩ facets (max-right, min-left) and two different ⟨111⟩ facets (max-left, min-right).

HAADF images for a NH at max and min tilt are shown in Fig. 4.14c and d, respectively, and indeed it can be seen that at max tilt, the left side has a high coverage while the right side is sparse, and at min tilt the inverse is true. To help visualize the coverage, line profiles from the max and min tilt (shown on Fig. 4.14e and f, respectively) are performed, and the coverage analyzed. The coverage is calculated by the percent of the line profile with HAADF intensity above a threshold, 20% of the maximum HAADF intensity is chosen qualitatively as giving the best representation of the nanoparticle coverage.

4.2 Magnetic and Plasmonic Nanocomposites

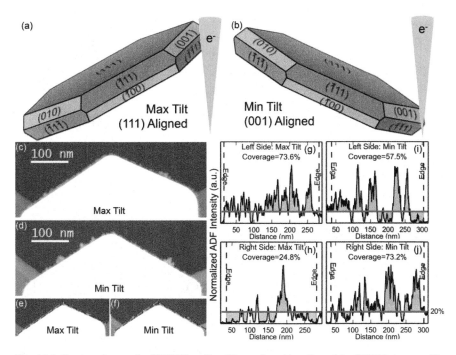

Fig. 4.14 Facet preference for SPION bonding. The preferred bonding of the SPIONs to a specific Au facet is determined through tilting the sample. (**a**) and (**b**) show two orientations of the NH with respect to the beam, termed max and min tilt, respectively, which align alternating facets with the electron beam. HAADF images of the max (**c**) and min (**d**) tilts show that on alternating facets there are varying concentrations of the SPIONs. Line profiles along the edges of the two tilts, (**e**)- max and (**f**)-min, are plotted, and the percent coverage is calculated by seeing where along the line profile the HAADF intensity is greater than a 20% threshold (red line). (**g**)–(**j**) Show the line profiles on the two different edges of the NH at the two tilts, and it can be seen that at max-left (**g**) and min-right (**j**) the coverage is significantly higher than max-right (**h**) and min-left (**i**). Due to the alternating nature of the facets, this demonstrates that SPIONs preferentially bond to one facet of the NH

In Fig. 4.14g–j, the four-line profiles are plotted, showing high coverage at max-left and min-right, and low coverages at max-right and min-left. It is worth noting that in the min-left profile (Fig. 4.14i) the coverage is artificially increased due to a large portion of the supporting carbon film that overlaps with the NH edge. Determining which crystal axis corresponds to the high-concentration facets is not possible from the STEM data, as the tilt limits of the STEM holders are ~20°, which is only enough to preferentially align a facet but not to tilt into a crystal zone access normal to the facet. As a result, the data is not sufficient to absolutely confirm that SPIONs selectively bond to the ⟨111⟩ and it does provide direct experimental evidence of preferential bonding on different gold facets in the Au NH-SPIONs as suggested in [39].

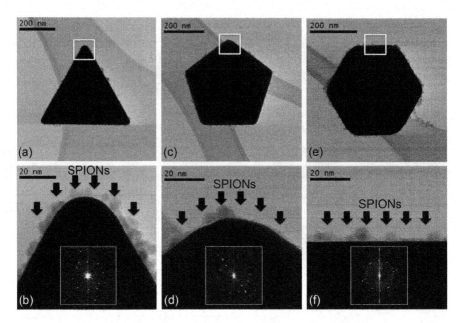

Fig. 4.15 Looking for epitaxial relationships between Au structures and SPIONs. (**a**) Bright field (BF) STEM image of a single Au NT-SPIONs nanostructure. (**b**) BF STEM image of NT tip outlined in (**a**) showing crystallinity of the SPIONs. The inset contains a fast Fourier transform (FFT) of the BF image and shows FFT spots arranged in a circular pattern indicating that the SPIONs are randomly oriented with respect to the NT. (**c**)–(**f**) BF STEM images of an Au icosahedra-SPIONs, (**c**) and (**d**), and an Au NH-SPIONs, (**e**) and (**f**), structure and the top corner/edge containing the SPIONs. In the inset of (**d**) and (**f**), the same circular FFT pattern is visible as in (**b**), indicating no epitaxial relationship between the Au and SPIONs occurs in any of the geometries present in the ensemble

It has been proposed that the mechanism for the increased bonding of the ⟨111⟩ Au facet is based in the PVP polymer that connects the SPIONs and the Au, not in the direct relationship between the SPIONs themselves and the Au [39]. To investigate the nature of the bonding, BF STEM imaging is performed. Figure 4.15 shows a NT, an icosahedron, and a NH with SPIONs and high-resolution images of the interface between the SPIONs and the Au substructures. It is seen that the Au NTs are of well-defined equilateral shape with their surface decorated by a monolayer of SPIONs as shown in Fig. 4.15a and b. The equivalent images of a SPION-decorated Au icosahedra and Au NH are shown in Fig. 4.15c through 4.15f. The resolution of the BF images in Fig. 4.15b, d, and f is high enough to see the lattice planes of the SPIONs, and the insets contain the FFT of their respective images. In the FFTs, each spot indicates a different crystal plane, and a circular ring pattern is an indication of many nonaligned crystal grains/nanoparticles. The circular FFT pattern is observed in both the NT and NH, indicating that the SPIONs are randomly oriented with respect to the Au NT, and no epitaxial relationship between the two exists.

4.2 Magnetic and Plasmonic Nanocomposites

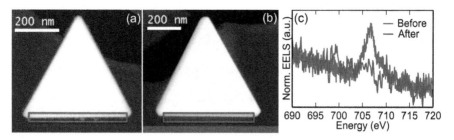

Fig. 4.16 Removing SPIONs post-synthesis. Despite being used to nucleate the gold nanostructures, the SPIONs can be removed post-synthesis through a dilute acidic treatment. (**a**) and (**b**) are HAADF images of nanotriangles from different samples, pre- and posttreatment. (**c**) Shows the Fe EELS L-edge from the bottom surfaces of both nanotriangles. No iron is detected in the posttreatment sample indicating that the SPIONs have been successfully removed

Since the only connection between the SPIONs and the Au substructure is through the PVP polymer, the SPIONs can be removed in a straightforward manner. Using diluted acidic solution an ensemble of the Au nanostructures can be obtained absent of the SPIONs that facilitated their nucleation. Figure 4.16 shows Au NT-SPIONs from an ensemble before and after the removal of the SPIONS (performed by Dr. Yu in Barcelona). Figure 4.16a shows a NT with the SPIONs still adhered to the surface, while Fig. 4.16b shows a NT without any SPIONs. The SPIONs are visible in the HAADF image presented in Fig. 4.16a, but none are observed in Fig. 4.16b. To further demonstrate the absence of iron from the treated Au nanostructures, the EELS core-loss data from the Fe-L_3 edge was collected from a large region at the bottom edge of both NTs. In the NT-SPIONs, the Fe-L_3 edge is easily observed but is completely absent from the NT posttreatment with HCl, indicating that SPIONs have been removed. The ability to remove SPIONs post-synthesis opens the future opportunity to study the plasmonic properties of the same planar gold nanostructures, with and without SPIONs.

4.2.3 The Optical Response of the Nanocomposites

The gold component of the composites is plasmonic and hence has a strong optical response that is highly sensitive to the morphology of the nanoparticle. Since the morphology of the gold has a strong impact on the plasmonic resonances, the control over the shape of the structures also results in the ability to tune the optical response.

Figure 4.17 shows the optical response of the nanoparticle ensembles under different synthesis conditions as determined by Dr. Yu and presented in [33]. Figure 4.17a and b shows how ensemble morphologies can vary significantly under different growth conditions, namely, the PVP in the solution during the addition of the HAuCl4 (a) and the temperature during the microwave heating stage (b). As can be seen in Fig. 4.17c and d, these changes in the basic morphologies of the

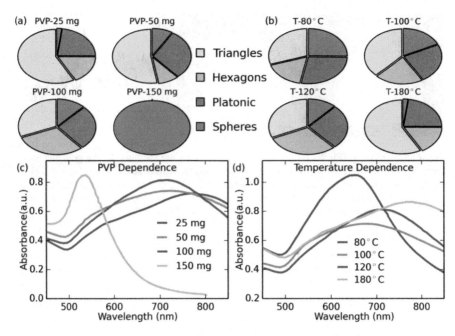

Fig. 4.17 Selectability of the optical response of nanocomposite ensembles. By changing the ratio of PVP to HAuCl$_4$ (**a**) and the temperature of microwave heating (**b**) during synthesis, the morphologies in the composite ensemble can be varied significantly. The optical response of the ensemble is plasmonic and hence dependent on morphologies and can be tuned across the visible regime by altering the PVP (**c**) or temperature (**d**)

nanoparticles vastly affect the optical response of the ensemble, allowing one to tune the peak response across a ~300 nm spectral range, by adjusting basic parameters of the synthesis process.

The response is due to the high sensitivity of SPRs to the morphology of the nanoparticle. Each of the different geometries supports different SPRs and each SPR is dependent on the size and local variations of the Au nanoparticle. By utilizing the low-loss portion of the EEL spectrum, with a Zeiss Libra200-MC, plasmonic resonances can be detected and mapped in the individual nanostructures. Figure 4.18 shows the plasmonic response of one of the nanotriangles.

Plasmonic analysis of the nanoparticle results in the detection of three distinct plasmon modes, the spatial profile of which is shown in Fig. 4.18a–c. Two of the modes are localized in the corners of the nanotriangle, while one is localized along the side: corner 1 at 1.6 eV (Fig. 4.18a), corner 2 at 2.1 eV (Fig. 4.18b), and the side mode at 2.4 eV (Fig. 4.18c). To better see the distinction between the modes (especially between the two modes at the corner, three different ROIs are examined at two different corners and one along the edge. ROI-1 (the top left corner) has its EEL spectrum plotted in Fig. 4.18d. From this plot it seems that there is only one peak at the corner. However, the peak here does not have the characteristic

4.2 Magnetic and Plasmonic Nanocomposites

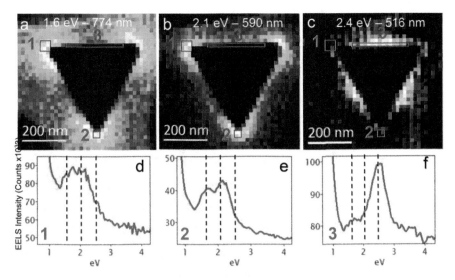

Fig. 4.18 Plasmon modes in the SPION-decorated nanotriangles. (**a**)–(**c**) The plasmon modes in an Au NT-SPION. The structure shows distinct plasmon modes at 1.6 eV (**a**), 2.1 eV (**b**), and 2.4 eV (**c**). Three ROIs are marked on the SIs labeled one through three, the EEL spectra for which are shown in (**d**)–(**f**). ROI-1 shows a broad peak at the corner (**d**), ROI-2 shows that the corner peak is actually composed of two smaller peaks at 1.6 and 2.1 eV (**e**), and ROI-3 shows that while the predominant mode on the edge is a side mode, there is also slight intensity from the 1.6-eV corner mode, indicating that this plasmon mode is the dipolar plasmon

Lorentzian shape of a plasmon mode. By examining another corner at the bottom of the nanotriangle with ROI-2, it can be seen why. Figure 4.18e shows that the peak from Fig. 4.18a really corresponds to the convolution of two modes, and here the EEL spectrum has a high enough signal to noise to clearly resolve the two Lorentzian-like peaks at 1.6 and 2.1 eV. Finally, by looking all along an edge (to accumulate the most signal), as is done in ROI-3, the side mode can be clearly visualized as an easily distinct peak.

Concerning the presence of the two different corner modes, it is important to note that there is a small peak at 1.6 eV along the edge (ROI-3) but no peak at 2.1 eV. Dipole modes in nanotriangles have been shown to have faint intensities along the edge [40] indicating that the 1.6 eV is the traditional dipole mode. The 2.1-eV corner mode could potentially be a higher-order multipolar mode due to the thickness of the nanostructures, which is higher than the thickness of most other nanotriangles reported in literature (100+ nm), but that is speculation and not well understood at this time.

By correlating the plasmon modes in Fig. 4.18 with the ensemble optical response in Fig. 4.17, the behavior of the ensemble can be correlated with the spatial profiles of different plasmon modes. For the 25-mg PVP, 180 °C ensemble that was composed mostly of nanotriangles, the response can actually be broken into three plasmon modes. There are two faint shoulders in the low- and high-500 nm

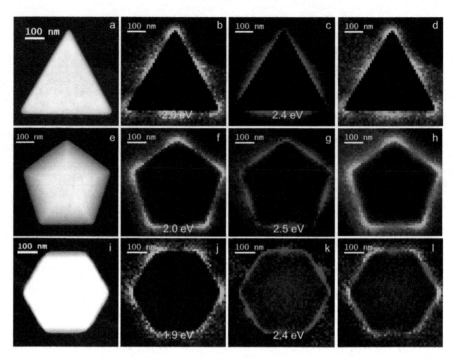

Fig. 4.19 Plasmon modes in varying geometries. Different geometries in the ensembles show the same basic behavior of corner modes at ~2 eV and side modes at ~2.4 eV. (**a**)–(**d**) The plasmonic response of an Au NT: (**a**) HAADF. (**b**) Corner mode – 2.0 eV. (**c**) Side mode – 2.4 eV. (**d**) Composite. (**e**)–(**h**) Corresponding images of (**a**)–(**d**) for nanoicosahedron, with corner mode at 2.0 eV and side mode at 2.5 eV. (**i**)–(**l**) Corresponding images of (**a**)–(**d**) for nanohexagon, with corner mode at 1.9 eV and side mode at 2.4 eV

range, and one large peak in the high-700 nm range, which are the same places that the distinct plasmon modes in Fig. 4.18 were observed, showing that the dominant plasmon mode in terms of the total optical response of the ensemble is likely coming from predominantly the side plasmon mode.

The plasmonic analysis can also be extended to the other geometries present in the ensembles. Figure 4.19 shows the plasmon profiles of the three predominant geometries found in the nanocomposite ensembles. Figure 4.19a–d shows the plasmon profiles for a nanotriangle (different than the one in Fig. 4.18). For this nanotriangle the splitting between the dipole and 2.1-eV corner plasmon mode was not well defined, so the convoluted peak is plotted in the SI, which results in some intensity being available on the edges of the NT. However, the same basic behavior of the plasmon modes observed in Fig. 4.18 is observed here with the corner modes centered at 2.0 eV and the side mode at 2.4 eV again. Figure 4.19e–h shows the same plasmonic analysis for a nanoicosahedron. The approximate size of the geometry is similar to the triangle, and the plasmon modes come out with the same general behavior: mode localized in the corner (at 2.0 eV) and a mode localized on the sides (at 2.5 eV). Finally, a nanohexagon is shown in Fig. 4.19i–l. The corner/side-mode localization is present here as well, with similar resonances for similar sizes. The data shows that for ensembles with different shapes, resonances in the same spectral regime can be achieved with varying spatial profiles on structures of different geometries.

The studies presented in this chapter present a variety of different STEM techniques for studying complex nanostructures. Using high-resolution imaging techniques and the associated image processing, in depth knowledge about the structural composition of nanostructures can be determined. Additionally, EELS provides a twofold benefit for nanoscale analysis between elemental mapping with core-loss EELS and probing optical excitations such as plasmons with low-loss EELS. The combination of the two makes STEM an ideal tool for the analysis of nanotechnology.

References

1. Shockley, W.: Transistor technology evokes new physics. Nobel Lecture, pp. 344–374 (1956)
2. Brattain, W.H., Garrett, C.: Surface properties of semiconductors. Physica **20**, 885–892 (1954)
3. Bardeen, J.: Great solid state physicists of the 20th century. Nobel Lecture, pp. 234–260 (2003)
4. Kilby, J.: Turning potential into reality: the invention of the integrated circuit. Nobel Lecture (2000)
5. Kamata, Y.: High-k/Ge MOSFETs for future nanoelectronics. Mater. Today **11**, 30–38 (2008)
6. Brunco, D., et al.: Germanium MOSFET devices: advances in materials understanding, process development, and electrical performance. J. Electrochem. Soc. **155**, H552–H561 (2008)
7. Kang, L., et al.: Electrical characteristics of highly reliable ultrathin hafnium oxide gate dielectric. IEEE Electron Device Lett. **21**, 181–183 (2000)
8. Lee, B.H., Kang, L., Nieh, R., Qi, W.J., Lee, J.C.: Thermal stability and electrical characteristics of ultrathin hafnium oxide gate dielectric reoxidized with rapid thermal annealing. Appl. Phys. Lett. **76**, 1926–1928 (2000)
9. Franco, J., et al.: 6Å EOT $Si_{0.45}Ge_{0.55}$ pMOSFET with optimized reliability (V_{DD}= 1V): meeting the NBTI lifetime target at ultra-thin EOT. In: Electron Devices Meeting (IEDM), 2010 IEEE International, pp. 4–11 (2010)
10. Mitard, J., et al.: 1mA/um-I_{ON} strained SiGe 45%-IFQW pFETs with raised and embedded S/D. In: 2011 Symposium on VLSI Circuits-Digest of Technical Papers (2011)
11. Tsetseris, L., Zhou, X., Fleetwood, D., Schrimpf, R., Pantelides, S.T.: Physical mechanisms of negative-bias temperature instability. Appl. Phys. Lett. **86**, 142103 (2005)
12. Grasser, T., Gos, W., Kaczer, B.: Dispersive transport and negative bias temperature instability: boundary conditions, initial conditions, and transport models. IEEE Trans. Device Mater. Reliab. **8**, 79–97 (2008)
13. Grasser, T., et al.: A two-stage model for negative bias temperature instability. In: 2009 IEEE International Reliability Physics Symposium, pp. 33–44 (2009)
14. Campbell, J.P., Lenahan, P.M., Krishnan, A.T., Krishnan, S.: Observations of NBTI-induced atomic-scale defects. IEEE Trans. Device Mater. Reliab. **6**, 117–122 (2006)
15. Ryan, J., Lenahan, P., Grasser, T., Enichlmair, H.: Recovery-free electron spin resonance observations of NBTI degradation. In: 2010 IEEE International Reliability Physics Symposium (IRPS), pp. 43–49 (2010)
16. Duan, G.X., et al.: Activation energies for oxide-and interface-trap charge generation due to negative-bias temperature stress of Si-capped SiGe-pMOSFETs. IEEE Trans. Device Mater. Reliab. **15**, 352–358 (2015)

17. Mamouni, F.E., et al.: Fin-width dependence of ionizing radiation-induced degradation in 100-nm gate length FinFETs. IEEE Trans. Nucl. Sci. **56**, 3250–3255 (2009)
18. Tsetseris, L., Zhou, X.J., Fleetwood, D.M., Schrimpf, R.D., Pantelides, S.T.: Hydrogen-related instabilities in MOS devices under bias temperature stress. IEEE Trans. Device Mater. Reliab. **7**, 502–508 (2007)
19. Hikavyy, A., et al.: SiGe SEG growth for buried channels p-MOS devices. ECS Trans. **25**, 201–210 (2009)
20. Tok, E., Ong, S., Kang, H.C.: Hydrogen desorption kinetics from the $Si_{1-x}Ge_x$ (100)-(2x1) surface. J. Chem. Phys. **120**, 5424–5431 (2004)
21. Huang, X. et al.: Sub 50-nm FinFET: PMOS. In: Electron Devices Meeting, 1999. IEDM'99. Technical Digest. International, pp. 67–70 (1999)
22. Singh, N., et al.: High-performance fully depleted silicon nanowire (diameter 5 nm) gate-all-around CMOS devices. IEEE Electron Device Lett. **27**, 383–386 (2006)
23. Gu, J., et al.: First experimental demonstration of gate-all-around III-V MOSFETs by top-down approach. In: 2011 IEEE International Electron Devices Meeting (IEDM), 33-2 (2011)
24. Shin, C.: State-of-the-art silicon device miniaturization technology and its challenges. IEICE Electron. Express **11**, 20142005 (2014)
25. Zhang, E.X., et al.: Total ionizing dose effects on strained Ge pMOS FinFETs on bulk Si. IEEE Trans. Nucl. Sci. **64**, 226–232 (2017)
26. Cooley, J.W., Tukey, J.W.: An algorithm for the machine calculation of complex Fourier series. Math. Comput. **19**, 297–301 (1965)
27. Dadgour, H.F., Endo, K., De, V.K., Banerjee, K.: Grain-orientation induced work function variation in nanoscale metal-gate transistors—Part I: modeling, analysis, and experimental validation. IEEE Trans. Electron Devices **57**, 2504–2514 (2010)
28. Matsukawa, T., et al.: Suppressing Vt and Gm variability of FinFETs using amorphous metal gates for 14 nm and beyond. In: Electron Devices Meeting (IEDM), 2012 IEEE International, 8-2 (2012)
29. Luke, G.P., Yeager, D., Emelianov, S.Y.: Biomedical applications of photoacoustic imaging with exogenous contrast agents. Ann. Biomed. Eng. **40**, 422–437 (2012)
30. Xu, L., et al.: Regiospecific plasmonic assemblies for in situ Raman spectroscopy in live cells. J. Am. Chem. Soc. **134**, 1699–1709 (2012)
31. Duncan, B., Kim, C., Rotello, V.M.: Gold nanoparticle platforms as drug and biomacromolecule delivery systems. J. Control. Release **148**, 122–127 (2010)
32. Hirsch, L., et al.: Nanoshell-mediated near-infrared thermal therapy of tumors under magnetic resonance guidance. Proc. Natl. Acad. Sci. **100**, 13549–13554 (2003)
33. Yu, S., et al.: Magnetic gold nanotriangles by microwave-assisted polyol synthesis. Nanoscale **7**, 14039–14046 (2015). ISSN:2040-3372
34. Hachtel, J.A., et al.: Gold nanotriangles decorated with superparamagnetic iron oxide nanoparticles. Faraday Discussions (2016)
35. Egerton, R.F.: Electron Energy-Loss Spectroscopy in the Electron Microscope, 3rd edn., 491 pp. Springer, Boston (2011). ISBN:1-4419-9582-X
36. Williams, D.B., Carter, C.B.: Transmission Electron Microscopy, pp. 3–17. Springer, Berlin (1996)
37. Egerton, R.F.: Electron energy-loss spectroscopy in the TEM. Rep. Prog. Phys. **72**, 016502 (2009). ISSN:0034-4885
38. Egerton, R.: Oscillator-strength parameterization of inner-shell cross sections. Ultramicroscopy **50**, 13–28 (1993)
39. Xia, Y., Xia, X., Peng, H.-C.: Shape-controlled synthesis of colloidal metal nanocrystals: thermodynamic versus kinetic products. J. Am. Chem. Soc. **137**, 7947–7966 (2015)
40. Geuquet, N., Henrard, L.: EELS and optical response of a noble metal nanoparticle in the frame of a discrete dipole approximation. Ultramicroscopy **110**, 1075–1080 (2010)

Chapter 5
Probing Plasmons in Three Dimensions

In the last chapter, I showed many different methods of characterizing samples at the nanoscale using the electron microscope and began to show the power of the STEM to characterize the optical response of nanostructures. Mapping plasmons is a useful process but is fairly straightforward to perform on modern STEMs and even SEMs. In this chapter I show how the type of plasmonic analysis shown in Sect. 4.2.3 can be taken to the next level by understanding the way that the STEM interacts with optical phenomena such as LSPRs. Here, I combine EELS and CL together to perform a joint analysis on a single plasmonic structure and show how the complementarity of the two can be exploited to access the three-dimensional plasmonic response of the system.

5.1 Plasmons in Three-Dimensional Structures

In recent years, an effort has been made to push past the two-dimensional planar structures that are easiest to study with electron microscopy and delve into structures with a three-dimensional plasmonic response. In order to access the full response of such nanostructures, many researchers have employed tilt series tomography to develop three-dimensional plasmonic maps of the spatial intensity distributions of individual plasmon modes [1–3] and to reconstruct the precise morphology of the structure and simulate its behavior [4, 5]. However, for complex geometries, where the 2D projection of the tilted system is difficult to interpret directly, tomography and computerized reconstructions can be prohibitively difficult or time-consuming. Thus, finding a way to use the standard STEM techniques to circumvent these limitations and developing a straightforward way to analyze surface plasmons in three dimensions are both desirable objectives.

By using the combination of spatially resolved EELS and CL in a STEM, three-dimensional experimental data for the surface plasmon modes of an individual nanoparticle can be accessed without reconstruction or simulation. The combination EELS/CL embodies a powerful complementarity that can be used to extract three-dimensional information about the plasmonic response not available from either spectroscopy individually, because EELS measures beam-induced electronic excitation, while CL measures radiative decay. Thus, in complex nanostructures with nonuniform dimensions, EELS is dominated by volume effects, while CL is dominated by surface effects. Here, the complementarity is exploited to analyze the plasmonic response of Ag nanoparticles decorating ZnO/MgO core/shell nanowires using a combined STEM-EELS and -CL analysis. The difference between EELS and CL yields additional information that permits the measurement and identification of distinct plasmon modes in all three dimensions directly from experimental data.

5.2 Complementary Spectroscopies in the Electron Microscope

The focus of this method is to demonstrate that the combination of the two spectroscopies, EELS and CL, provides information about the surface plasmon modes of a single nanoparticle that cannot be accessed by either technique individually. Here, the analysis of a single nanoparticle is presented to demonstrate that by the use of joint EELS/CL the three-dimensional analysis of individual plasmonic elements in a complex multiple-element system can be obtained.

The ZnO nanowires were grown by vapor-solid deposition by Claire Marvinney and Daniel Mayo at Fisk University. Glancing angle deposition by electron-beam evaporation was then used to deposit an MgO spacer layer and to decorate the surface with Ag nanoparticles [6]. The nanoparticles on the nanowire have characteristic lateral dimensions ranging from 2 to 5 nm to more than 100 nm. Most nanoparticles are roughly hemispherical in shape, but there is no predictable symmetry or orientation with respect to the nanowire axis. The ZnO nanowire itself is optically active and a strong emitter, so I chose a nanowire with a thick MgO shell (~70 nm) to ensure that the nanoparticle is insulated from the wire and isolated from hot-electron transfer effects and plasmon-exciton interaction with the ZnO, which are strongest in the <30 nm spacer range and almost absent at 70 nm spacer thicknesses [7]. Many nanoparticles were studied, however one with high-aspect ratios was chosen for analysis, as the resultant splitting of the dominant plasmon modes along the different principal axes is necessary to demonstrate the acquisition of three-dimensional plasmonic data.

Plasmonic excitations are detected and mapped in different ways in EELS and CL. In plasmon maps generated by both techniques, the spatial location of the plasmon intensity is determined by position of the probe when the fast electrons in the beam scatter inelastically from the conduction band electrons of the metal [8–11]. However, the plasmon signature measured by EELS is the energy lost during transmission through the sample, which directly relates to the combined sum of the energies for all the excitations generated by each beam electron during transmission. As a result, EELS is sensitive to all plasmons, including "dark," high-angular-momentum plasmon modes that cannot decay radiatively and high-energy plasmon modes that decay outside the visible spectrum [12–14]. On the other hand, CL spectra incorporate signals from the radiative decay of the beam-induced excitations. This limits CL to electronic excitations that effectively out-couple to photons in the detectable range of the spectrometer, thereby directly yielding information about the efficiency of individual plasmon absorption or emission modes [12, 15–17]. As a result, the detected plasmonic response can differ greatly when comparing EELS to CL even in the same system, especially in complex geometries with varying thicknesses. By assessing the physics underlying these differences The system can be understood more thoroughly.

The crux of the current experiment is the determination of the spatial distribution of plasmon response through the construction of a SI of the nanoparticle and the difference between the SIs collected in STEM-EELS and CL. In both spectroscopies, a spectrum is collected at each x–y point in a two-dimensional scan: in EELS from the energy loss experienced by the beam electrons and in CL from the radiative decay of the beam-induced excitations. When the spatial scan is complete, one obtains a 3D data set with two spatial dimensions and one energy dimension. To form 2D images, slices are taken across the energy dimension, and the integrated intensity over a given spectral range is converted into the pixel intensity, allowing the mapping of specific optical features.

It is important to note that the detection efficiency is significantly higher in EELS than in CL. As a result there are two principal differences between the SI acquisition methods in EELS and CL: pixel time and beam current. In terms of the pixel time, for EELS, all SI are taken with 0.2 s per pixel, while in CL, 20 s per pixel is needed for modest signal to noise. In terms of the beam current, EELS measurements use a low beam current (less than 20 pA) that is used to optimize the EELS signal and maximize spatial resolution, while for CL, a high beam current is used, ~2 nA, in order to maximize CL signal intensity. As a result of the high beam current and long acquisition times in CL, beam-induced damage becomes a significant concern, so the pixel size is increased and sub-pixel scanning is applied to reduce the total dose at each point, resulting in significantly lower spatial resolution in CL SIs compared to EELS. To assure that the beam has not altered the nanoparticle significantly, the plasmon resonances observed in the CL-SI are, as with the EELS, double checked after acquisition and confirmed to retain approximately the same peak position, amplitude and width.

To determine the experimental energy values for the plasmon peaks, representative spectra are taken from the SI in regions of peak intensity of each plasmon mode in each spectroscopy. The spectra are then fit with a nonlinear least squares regression to determine the peak positions.

5.2.1 Surface Plasmons Observed in Both EELS and CL

Differences between the way plasmons are detected in EELS and CL result in some peaks being visible in both spectroscopies, while some are only observed in one or the other. In Fig. 5.1 a single random-morphology Ag nanoparticle on the surface of the nanowire is analyzed, and the plasmonic maps in EELS and CL for the plasmon modes that appear in both spectroscopies are compared. An ADF image of the nanoparticle is shown in Fig. 5.1a, and the particle's plasmonic response is mapped through SIs in Fig. 3.1b–e, with the representative spectra of the plasmon modes being shown in Fig. 5.1f, g. Peaks near 2.0 eV (centered at 2.03 eV in EELS and 1.95 eV in CL) and 3.0 eV (2.87 eV EELS, 3.07 eV CL) appear both in EELS and CL. By comparing the SI slices for these peaks in EELS (Fig. 5.1b, d) to the CL slices (Fig. 5.1c, e), the higher detection efficiency and spatial resolution of EELS can be exploited to identify the nature of the resonance. The 2.0-eV plasmon is localized at the top and bottom of the nanoparticle, while the 3.0-eV peak is

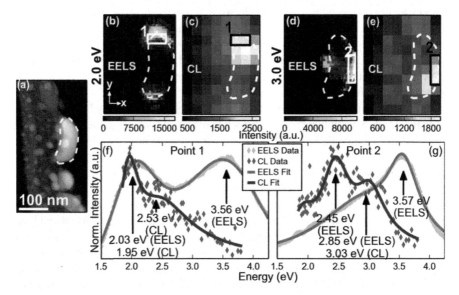

Fig. 5.1 Electron-energy loss spectroscopy vs. cathodoluminescence. (**a**) An ADF image of an Ag nanoparticle on the surface of an insulating nanowire. (**b**)–(**e**) SIs of the plasmon modes present in both EELS and CL. (**b**) and (**c**) show the SIs for EELS and CL, respectively, of a long-axis (LA) plasmon mode oscillating from the top to the bottom (along the y-axis) at 2.0-eV that is more strongly excited at the top of the nanoparticle. (**d**) EEL and (**e**) CL-SI, respectively, also show a short-axis (SA) plasmon mode (oscillating along the x-axis) at 3.0-eV that is stronger on the right side of the nanoparticle. Representative spectra are taken from regions of maximum intensity of the two spectra. Point 1 at the top of the nanoparticle (**f**) and Point 2 at the right side (**g**) show that while both EELS and CL show peaks at 2.0-eV and 3.0-eV corresponding to the LA and SA plasmons, there is a 2.5-eV peak present in CL and not EELS, as well as a 3.6-eV peak present in EELS and not CL

localized along the sides, indicating that the nanoparticle exhibits the characteristic response of an ellipsoidal nanoparticle; the longitudinal dipole surface plasmons are split into top-to-bottom, long-axis (LA) and side-to-side, short-axis (SA) modes.

It is worth noting that both extrema of the plasmonic resonance are observed in EELS, but not in CL, probably due to the lower signal-to-noise ratio in CL. There is still CL intensity at the bottom and left-hand side of the nanoparticle, but the resonance is too weak to be sufficiently distinguished from the background noise. There are two likely root causes for the asymmetry in the plasmon intensities at the nanoparticle edges: a thickness variation in the nanoparticle and the presence of the MgO spacer. The discrepancy in intensity between the top and the bottom of the nanoparticle seems to be a result of the nonuniform thickness. The particle is thicker on the top than the bottom, resulting in increased signal in both spectroscopies on the top side of the nanoparticle. The thickness measurements and discussion surrounding them are treated in more detail later in the text. The MgO spacer affects all plasmon modes by altering the dielectric environment of the nanoparticle. However, the SA mode, especially when excited from the left side, is particularly affected and suffers a reduced intensity due to the presence of the MgO, which damps the electric field of the plasmon resonance and reduces the electromagnetic coupling to the electron beam.

5.2.2 Constant Background Subtraction in EELS Spectrum Imaging

To accurately display the locations of the various plasmon peaks in EELS, a constant background subtraction is used. Normally, in an EELS low-loss plasmon experiment, the background subtraction method for spectrum imaging is removing the ZLP and fitting the resulting background-free spectra. However, in these experiments two different SIs are acquired, one that contains the ZLP and one which does not. The SI containing the ZLP is required for the low-loss Fourier-log thickness measurements, which will be discussed in Sect. 5.3.1. However, the ZLP is such a strong feature in the EELS spectrum that short pixel acquisition times are required to prevent damage to the EEL spectrometer. To circumvent this limitation, a second SI is acquired over a spectral range that blocks out the majority of the ZLP but includes the low-loss region, allowing for longer pixel acquisition times without damaging the spectrometer and improving signal to noise. Without the entire ZLP in the spectrum, the ZLP cannot be extracted rigorously and consistently, so that the traditional method of background subtraction cannot be used. The benefits of blocking the ZLP can be seen in Fig. 5.2, which shows spectra from the identical spot from the two SIs. First, where ZLP is included and extracted through the reflected tail method, Fig. 5.2a, and the SI where the ZLP is blocked off the recorded spectrum, Fig. 5.2b. The improvement in signal to noise between the two methods is significant and clearly evident.

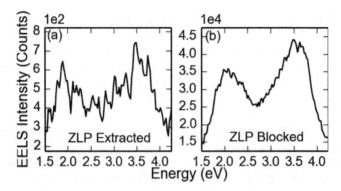

Fig. 5.2 Different methods of background subtraction. (**a**) Low acquisition time including the ZLP which is then extracted through the reflected tail method. (**b**) High acquisition time blocking out the ZLP then subtracting a power-law background. The general shape of the spectra is consistent between the two methods; however, the signal to noise is greatly improved in (**b**) allowing for more accurate measurement of peak position

The constant background subtraction method is demonstrated in Fig. 5.3 where the spectra from the top end of the nanoparticle, where the LA plasmon is highly active, and from the middle of the nanowire, where there is little to no plasmonic activity. A comparison of the two spectra shows the effect of the ZLP tail, which results in a strong total intensity present at both locations. In Fig. 5.3a the spectra with the LA plasmon is plotted. The hatched region shows the total intensity, while the highlighted portion of the hatched region shows where the constant background subtraction identifies plasmonic intensity. In comparison, Fig. 5.3b still shows significant total intensity due to the tail of the ZLP, but no clear plasmon peak. As a result the constant background level is higher than the EELS signal, and no plasmonic intensity is identified. Figure 5.3c shows an ADF image of the nanoparticle as reference for the SIs obtained using the total intensity (Fig. 5.3d) and the constant background-subtracted intensity (Fig. 5.3e). From (d) it is still possible to see that at ~2.0 eV there is some increased plasmonic intensity at the top and bottom of the nanoparticle with respect to the remainder of the region of interest but (e) shows the location of the peak plasmonic intensities much more clearly.

As a result of the constant background subtraction, the EEL spectra have a true zero level corresponding to the point at which the subtracted background exceeds the EELS signal at the plasmon resonance frequency. The CL signal, on the other hand, contains no corresponding background artifact, so no real zero level occurs, and background noise may appear in the luminescence spectra.

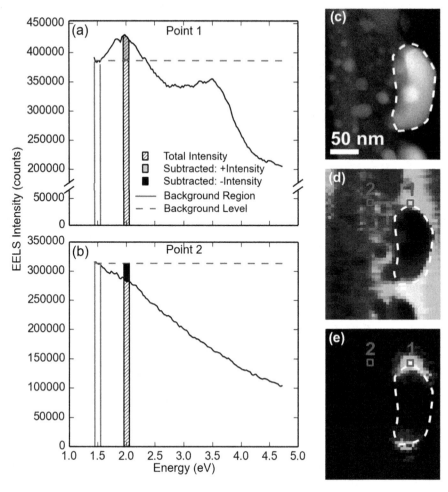

Fig. 5.3 Constant background subtraction. Spectra show how the constant background subtraction is used to isolate plasmonic signal in an EELS spectra. (**a**) The EEL spectrum from Point 1, both total intensity and the subtracted signal show strong intensity here. (**b**) The EEL spectrum from Point 2. No plasmon is present; the total intensity still shows signal due to the tail of the zero loss peak, but the constant background subtraction correctly finds no plasmonic activity. (**c**)–(**e**) The ADF image of the nanoparticle (**c**) and the SIs using total intensity (**d**) or the constant background-subtracted intensity (**e**). With the ZLP blocked, a power-law background can be fit to the tail for individual EEL spectra and look at the background-subtracted EELS data. Because of the varying thicknesses and compositions in the sample, no single method of power-law background subtraction and peak fitting could be found that would work uniformly across all spectra in the SI. As a result, while individual spectra show the power-law background-subtracted signal, the constant background subtraction method was used for the SI

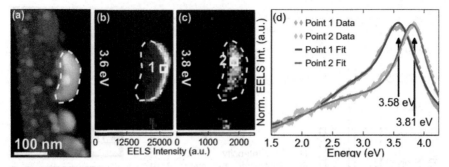

Fig. 5.4 Plasmon modes observed only in EELS. (**a**) ADF image of a nanoparticle. (**b**), (**c**) SIs of the two plasmon modes only observed in EELS at 3.6 eV (**b**) and 3.8 eV (**c**). (**d**) The EEL spectra from points 1 and 2 marked on (**b**) and (**c**). The two peaks refer to the transverse and bulk plasmons modes in Ag [18, 19]. The SI shows that the bulk plasmon is localized in the body of the nanoparticle and the transverse plasmon to the nanoparticle surface. The transverse mode appears to only be strong on the right side of the nanoparticle, but it is present along the entire perimeter of the nanoparticle, with a significantly reduced intensity. The two modes are visible in EELS but not in CL because of their ultraviolet energies that are outside the detection range of the CL spectrometer, but their presence in the EELS spectra are important for joint CL/EELS analysis

5.2.3 Surface Plasmons Observed Only in EELS

In Fig. 5.1f, g, peaks at 3.56 and 3.57 eV, respectively, were observed in EELS, with no corresponding feature in CL. The data also reveal another such peak at 3.78 eV. Figure 5.4 shows an ADF image of the nanoparticle (Fig. 5.4a), the SIs for each mode (Fig. 5.4b, c), and the spectra corresponding to each (Fig. 5.4d). The nature of both of these peaks is well known from previous research. They correspond to the transverse surface plasmon at 3.6 eV (Fig. 5.4b) and the bulk plasmon at 3.8 eV (Fig. 5.4c) of Ag [18, 19]. It is important to note that neither mode is geometry dependent. The bulk plasmon would be observed in any Ag sample, and the transverse plasmon is generally present at the surface of any Ag nanoparticle [20].

It is important to note that the transverse mode can be detected around the entire perimeter of the nanoparticle even where little to no intensity is visible in the SI. The differences in intensity for the transverse plasmon SI arise because the electron beam is transmitted through a thick MgO layer on one side of the nanoparticle, but only through the thin supporting carbon grid on the other, which significantly affects the total detected intensity. The wavelengths corresponding to these plasmons, detected by EELS, are 350 nm (transverse) and 325 nm (bulk), respectively, too far into the ultraviolet to be efficiently detected in the CL spectrometer available in this experiment. The detectable range of EELS is not limited to the visible regime, and thus EELS can detect the higher-energy ultraviolet modes. However, because they are undetectable in CL, the combined CL/EELS analysis cannot be for these modes. Their presence in the EELS spectra, however, will help in the combined CL/EELS analysis of the other plasmon modes, as will be shown later.

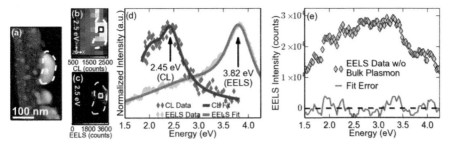

Fig. 5.5 Plasmon modes observed only in CL. (**a**) ADF image of nanoparticle. (**b**) CL-SI of the 2.5-eV peak observed in CL but not in EELS. (**c**) EEL-SI at 2.5 eV. Note that intensity is only observed in the SI in regions where the 3.0-eV SA plasmon (shown in Fig. 5.1d) has high intensity, indicating that the 2.5-eV EEL-SI is mainly due to the tail of this plasmon. (**d**) CL and EELS spectra from the center of the nanoparticle. Here the EELS spectrum is dominated by the bulk plasmon. Since EELS is a measure of excitation, the signal over thick regions is dominated by bulk features. However, CL as a measure of decay is dominated by surface effects, demonstrating that the 2.5-eV feature is a surface effect localized above the bulk of the nanoparticle. (**e**) The difference between EELS data and fit. The EELS fit has two peaks, one sharp peak corresponding to the bulk plasmon and a weak and broad plasmon peak centered at around 2.75 eV. The peak is too broad to be a single plasmon and demonstrates that EELS detects signal in the same region where the 2.5-eV peak in CL is observed but that it is too weak to be resolved into individual peaks. The behavior of the peak is consistent with an out-of-plane (OOP) surface plasmon, localized over the thickest part of the nanoparticle and hence unresolvable in EELS, but completely detectable in CL, exemplifying the power of the complementary EELS/CL analysis

5.2.4 Surface Plasmons Observed Only in CL

Finally, there is the feature present in CL but not in EELS. Figure 5.5, as well as both spectra in Fig. 5.1, shows a CL peak near 2.5 eV that has no corresponding EELS peak. The nanoparticle is shown in Fig. 5.5a, and the CL-SI of the plasmon mode is shown in Fig. 5.5b. Comparing to the previous CL-SI, it is evident that the 2.5-eV peak has a unique spatial localization, completely distinct from the LA and SA plasmon modes shown in Fig. 5.1c, e, indicating that it must be a separate plasmonic feature. The EEL and CL spectra from the center of the nanoparticle, shown in Fig. 5.5c, demonstrate that the 2.5-eV CL peak (centered here at 2.48 eV) is not observed in EELS, where only the bulk plasmon is detected. Figure 5.5c shows an EEL-SI at 2.5 eV, and while some intensity can be seen in the SI, those same locations correspond to the high-intensity regions of the SA plasmon in Fig. 5.1d. Due to the constant background subtraction method employed in the SI, the tail of a dominant plasmon can result in detected signal in the SI. As a result, the signal detected in the EELS at 2.5 eV is more likely to be due to the tail of the short-axis plasmon rather than a unique 2.5 eV plasmon.

In order to understand why the peak is present in one spectroscopy but not the other, the core distinction between EELS and CL is reexamined, excitation vs. radiative decay. One of the most important consequences of the distinction is how the signals of the two spectroscopies react to multiple electron-sample interactions. First, the total EELS intensity across the entire spectrum decreases when the electron beam interacts with the bulk, since a large portion of the electrons are elastically scattered to high angles and do not enter the spectrometer. Furthermore, EELS requires single-interaction events for the measured energy loss to be directly interpretable. If the beam interacts multiple times with the sample during transmission, the measured energy loss has contributions from each interaction, and the true values of each energy-loss event are obscured. Thus, even though the total plasmonic excitation may be stronger in a thick part of the sample, the detected plasmonic EELS signal is strongest, and most directly interpretable, along the edge of the nanoparticle. At that spatial location, the electron beam couples via the evanescent plasmon field and excites the plasmon nonlocally without interacting with the bulk of the nanoparticle, allowing for efficient single-interaction plasmon excitations [12].

In contrast to EELS, multiple interactions and elastic scattering do not attenuate CL signal. Surface plasmons and other radiative decay events are captured regardless of subsequent beam/sample interactions, because the emitted photons, rather than the transmitted electrons, are detected in CL. In fact, bulk luminescence effects can be attenuated in CL as the emitted photons are subject to reabsorption in the low-skin-depth metals, while surface luminescence (such as that from surface plasmons) is efficiently captured.

In STEM, or any other electron beam technique, the strength of the plasmonic excitation from the electron beam is determined by the strongest inelastic scattering from the surface plasmon mode. However, the strength of the detected EELS intensity is determined by the area of most-efficient, single-interaction excitation, while the detected CL intensity is determined by the strongest plasmonic excitation of the observed mode. So the strongest CL intensity should occur at the thickest part of the nanoparticle where the studied plasmon mode is active (i.e., on the top and bottom of the nanoparticle for the LA mode). Recall from Figs. 5.1 and 5.4 that the EEL-SIs show the strongest plasmon intensity outside the boundaries of the nanoparticle, while the CL-SIs show the strongest intensity within the boundaries.

The distinction between the way EELS and CL detect plasmons explains why the 2.5-eV plasmon is not observed in EELS. By examining the spectrum in Fig. 5.5d more closely a low-amplitude, broad peak on the tail of the bulk plasmon signal can be observed. The data are fit with two Lorentzians, one for the bulk plasmon and one for the weak tail of the plasmon peak. Figure 5.5e shows the EELS data with the fitted bulk plasmon peak subtracted. The peak has a full width at half maximum of 1.27 eV which is 30% broader than any other peak fit in the data set; the full parameters of all the fits are included in Appendix B. Due to this large width, it is likely that the peak arises from multiple plasmonic features, including the 2.5-eV plasmon being excited in the center of the nanoparticle. However, since EELS requires efficient single-interaction excitations to accurately account for a peak in the signal, the spectrum is dominated by the bulk plasmon in this region, and the other plasmonic peaks cannot be resolved individually.

From the inherent characteristics of the two spectroscopies, it can be said that when the electron beam is transmitting through a thick sample, the EELS signal only shows bulk effects strongly, while the CL can still show the surface effects. Since the 2.5-eV peak is strongly detected in CL but is unresolvable as an individual peak in EELS, the peak can be determined to be a surface effect localized above the bulk of the nanoparticle. The feature is likely plasmonic, as the CL spectra shown in Fig. 5.1 demonstrate it and has an amplitude and width comparable to the LA and SA modes. Longitudinal, out-of-plane (OOP) plasmons have been shown to be strongly active in CL, and such a mode would exhibit the type of localization observed in the CL-SI [16, 17, 21]. The supporting evidence indicates that the 2.5-eV peak is an OOP plasmon mode, and with the identification of the OOP mode, the characteristic frequency, spatial distribution, and location of peak intensity for the dominant longitudinal plasmon modes in all three dimensions have been determined purely experimentally without reconstruction or simulation.

5.3 Validation of Experimental Results

5.3.1 Approximating Nanoparticle Geometries

To confirm the preceding experimental analysis of the plasmon peaks in the random morphology nanoparticle, FDTD simulations were compared to the spectroscopic data. The morphology of the nanoparticle is not precisely known, and hence direct simulation of the exact plasmonic response is not possible. However, approximating the morphology of the nanoparticle in a simpler geometry and comparing simulations to the experimental results can function as validation of the experimental analysis by demonstrating that the simplified geometry, where the plasmonic response is known exactly, behaves in a similar way to the unknown geometry, where the plasmonic response is determined experimentally.

In order to reasonably approximate the size and shape of the nanoparticle and simulate its plasmonic response, information about the nanoparticle geometry that has already been established is reexamined. A half-ellipsoid on an MgO substrate is chosen to represent the nanoparticle for two reasons. Firstly, the shape of the genuine nanoparticle is likely quasi-ellipsoidal. The splitting of the LA and SA modes is a common feature in ellipsoidal nanoparticles, and the splitting observed in this experiment is similar to those established in literature [22]. Secondly, a half-ellipsoid is chosen as opposed to a full ellipsoid since the MgO shell, on which the Ag particle is deposited, is flat.

The x and y dimensions of the half-ellipsoid are measured from the calibrated STEM image shown in Fig. 5.6a. For the z-dimension, the EEL-SI can be used to determine the thickness of the nanoparticle. In the log-ratio (LR) technique, the thickness is determined from an EELS equation by the following equation, $t/\lambda = \ln(I_{\text{Spec}}/I_{\text{ZLP}})$, where t is the thickness, λ is the inelastic mean free path length, I_{Spec} is the integrated intensity of the EEL spectrum without the ZLP, and I_{ZLP} is the integrated intensity of the ZLP. It is acquired over a spectral range up to 46 eV, due to

Fig. 5.6 EELS thickness measurements. (**a**) ADF image of the nanoparticle with the measured width (~70 nm) and length (~140 nm) of the nanoparticle. (**b**) Thickness map of nanoparticle/nanowire assembly in terms of inelastic scattering mean free path lengths (λ). Nanoparticle is measured to be approximately 1 λ thick corresponding to a physical thickness of 100 nm [24]

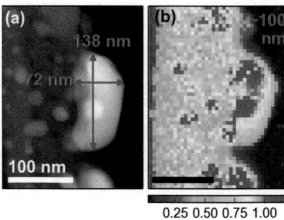

the small energy dispersion used for acquisition. Conventionally, a range of 200 eV, or preferably even higher, is used to determine thickness measurements. A power-law tail was added to simulate the unaccounted for portion of the EEL spectrum; however, this introduces further error into a calculation that already has an expected error of ±20% [23]. Additionally, different methods of calculating the λ result in different values, and microscope parameters (such as the convergence and collection angles, along with the operating voltage) have been found to significantly affect λ. The measurements are taken with a convergence angle of 30 mrad, a collection angle of 36 mrad, and an operating voltage of 200 kV. Using these parameters, I used two different methods of determining λ, the method proposed by Yang and Egerton (75 nm) [23], and the values determined from Iakoubovskii et al. (100 nm) [24]. The values from Yang and Egerton have been known to underestimate λ for microscopes with a collection angle over 10 mrad, so I choose the value from Iakoubovskii et al., for the accepted value.

The LR method determines the exact relative thickness in units of λ, but measurement of λ has a known error of approximately 20%, due to inhomogeneities and defects in the material alongside electron beam diffraction effects which can affect the EELS intensity [23, 25, 26]. The thickness map of the nanowire/nanoparticle assembly is generated by performing the LR measurement on each spectrum in the SI and generates a 2D relative thickness map (Fig. 5.6b). It can be seen that the thickness is roughly equal to 1 λ, corresponding to a physical thickness of approximately 100 nm for Ag [24].

Additionally, Fig. 5.6b shows that the depth at the top of the nanoparticle is greater than at the bottom, which offers a possible explanation as to the localization of the LA plasmon mode at the top of the nanoparticle in Fig. 5.1b. Since the nanoparticle is thicker at the top than the bottom, the inelastic scattering cross section for the plasmon should be higher at the top than at the bottom, and the detected plasmonic intensity should be stronger. The SA mode does not exhibit the same increased excitation at the top of the nanoparticle with respect to the bottom, as one might expect. However, this highlights the importance of a purely experimental analysis for unknown geometries, as morphological variations such as curvature or sharp edges can also affect the plasmonic intensity significantly [27–29]. Since the morphology of the nanoparticle is not precisely known, the reason why the SA mode does not exhibit increased intensity in the thicker region cannot be stated absolutely or even that the increased LA mode intensity at the top of the nanoparticle is due to the increased thickness. All that can be concluded is that the detected EELS and CL signals for the LA plasmon are greater at the top of the nanoparticle than at the bottom and that the increased thickness is a likely cause.

5.3.2 Finite-Difference Time-Domain Confirmation of Experimental Analysis

With the approximate dimensions of the nanoparticle determined, simulations of the plasmonic response of the nanoparticle become credible. Lumerical FDTD Solutions^{TM} code was by Claire Marvinney at Vanderbilt University used to solve the Maxwell equations for each orientation of the three principal axes of the nanoparticle relative to the polarization of the exciting plane wave. The simulation was for a half-ellipsoid Ag nanoparticle on a 170 nm × 170 nm × 80 nm MgO substrate. Additionally, the samples were not kept in vacuum, so a 1.6 nm shell of AgS_2 of 1.6 nm was included to simulate the experimentally measured effect of Ag tarnishing in atmosphere [30]. The dimensions of the nanoparticle were determined from the STEM measurements in Sect. 3.3.1, to be 140 nm × 100 nm × 70 nm. The nanoparticle and the MgO film it was on were rotated together to allow each polarization to be measured. The full parameters of the simulation are as follows. Simulated region: x, 300 nm; y, 300 nm; z, 800 nm; boundary conditions, periodic. Source: shape, plane wave; injection axis, z-axis; direction, backward; polarization, x-polarization, angle = 0°, simulation run three times for all three rotations of nanoparticle—y-polarization, angle = 90°, simulation run three times for all three rotations of nanoparticle; wavelength range, 200–1200 nm. Mesh region: location, centered on particle; size, $x = y = z = 170$ nm; max step, $dx = dy = dz = 2$ nm. Detectors: type, 2D; number, four; location, three cutting each Cartesian plane through the center of the particle (x–y, y–z, z–x) for electric field maps, one at the bottom of the simulation (source at top, x–y plane detector at bottom) to examine transmission spectrum; recorded data, in-plane detectors – standard Fourier transform, electric field, magnetic field, power – bottom detector – standard Fourier transform, electric field, Poynting vector, and power.

In Fig. 5.7, Ag half-ellipsoids on an MgO substrates are analyzed with FDTD. The plasmonic response of the system to plane waves polarized along the major axes of the ellipsoid is calculated. The resulting transmission data of a nanoparticle with dimensions determined in the previous section, 140 nm × 100 nm × 70 nm, are

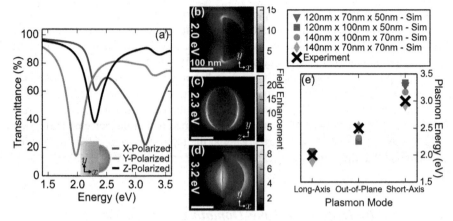

Fig. 5.7 Validating experimental analysis with simulation. The plasmonic response of a half-ellipsoid Ag nanoparticle on a MgO substrate with the dimensions determined by STEM-analysis (shown in Fig. 5.6) is determined through finite-difference time-domain calculations. (**a**) Transmission data from plane waves polarized at different orientations to the main axes of the nanoparticle. Three plasmon peaks are observed at 2.0, 2.3, and 3.2 eV. (**b**)–(**d**) From the resulting plasmonic field enhancement maps, it is shown that the modes correspond to (**b**) a 2.0-eV LA mode, (**c**) a 2.3-eV OOP mode, and (**d**) a 3.2-eV SA mode. Given that the precise morphology of the nanoparticle is not known, simulations are performed on multiple variations of the nanoparticle size in all three dimensions. (**e**) The results of these simulations are compared to the values for the three plasmon modes determined experimentally, and it can be seen that the experimental values fall within the predicted range determined from the simulations. The agreement indicates that the three-dimensional plasmon response detected, mapped, and identified from a purely experimental analysis without tomographic reconstructions is consistent with the known three-dimensional response of a nanoparticle with a simplified, but similar, geometry

shown in Fig. 5.7a. The plot exhibits three distinct plasmon peaks appearing at 1.98, 2.32, and 3.17 eV. By plotting the electric field enhancement of the plasmon modes of each of the resonances (Fig. 5.7b–d), it can be seen that the peaks correspond to LA (1.98 eV), SA (3.17 eV), and OOP (2.32 eV) plasmon modes.

Given that the thickness measurements and the nanoparticle geometry are approximate, the three plasmon modes are also calculated for half-ellipsoids of slightly different sizes to establish a spectral range in which each of the plasmons appear. The three other half-ellipsoid dimensions are 120 nm × 70 nm × 50 nm, 120 nm × 100 nm × 50 nm, and 140 nm × 70 nm × 70 nm. The plasmons corresponding to the half-ellipsoids are LA-1.97 eV/OOP-2.49 eV/SA-3.29 eV, LA-2.07 eV/OOP-2.26 eV/SA-3.34 eV, and LA-1.88 eV/OOP-2.54 eV/SA-2.92 eV. The simulated and experimental values are plotted together in Fig. 5.7e. It can be seen that the experimental values of 2.0, 3.0, and 2.5 eV fall entirely within the range of values established by simulations.

The size variations in the simulated nanoparticles are significantly larger than the predicted error in the experimental measurements of the nanoparticle size, because the precise morphology of the nanoparticle is not known, and morphological differences can account for significant variations in the plasmon mode energies as well. A larger range of half-ellipsoid dimensions is taken to establish a larger range over which the various plasmon modes can deviate under geometric variance. The STEM-images, EELS data, and thickness map show that while the nanoparticle is not a perfect half-ellipsoid, the geometry is reasonably similar, and the plasmon modes should be similar as well. The agreement between experimental data and the theoretical predictions is not a direct comparison of the two plasmonic models, as much as a sanity check that the experimental analysis is yielding physically reasonable results for a quasi-ellipsoidal structure. The experimental modes falling directly into the range established by the simulations indicate that the experimental joint EELS/CL analysis of the dominant nanoparticle plasmon modes is consistent with the plasmonic behavior of a nanoparticle of similar geometry. Furthermore, the simulations confirm that an experimental analysis based on the complementary physical origin of the two spectroscopic signals, without simulations, on a nanoparticle with an unknown morphology gives information on the plasmonic response of complex nanostructures. The amount of information obtained in this phenomenological approach is comparable to what is obtainable from an analysis of nanoparticles with predetermined morphologies, undergirded by electromagnetic simulations.

References

1. Nicoletti, O., et al.: Three-dimensional imaging of localized surface plasmon resonances of metal nanoparticles. Nature **502**, 80–84 (2013). ISSN:0028-0836
2. Li, G., et al.: Spatially mapping energy transfer from single plasmonic particles to semiconductor substrates via STEM/EELS. Nano Lett. **15**. WOS:000354906000103, 3465–3471 (2015). ISSN:1530-6984
3. Atre, A.C., et al.: Nanoscale optical tomography with cathodoluminescence spectroscopy. Nat. Nano **10**, 429–436. ISSN:1748-3387 (2015)
4. Perassi, E.M., et al.: Using highly accurate 3D nanometrology to model the optical properties of highly irregular nanoparticles: a powerful tool for rational design of plasmonic devices. Nano Lett. **10**, 2097–2104 (2010). ISSN:1530-6984
5. Hernández-Garrido, J.C., et al.: Exploring the benefits of electron tomography to characterize the precise morphology of core–shell Au@Ag nanoparticles and its implications on their plasmonic properties. Nanoscale **6**, 12696–12702 (2014). ISSN:2040-3372
6. Mayo, D.C., et al.: Surface-plasmon mediated photoluminescence from Ag-coated ZnO/MgO core-shell nanowires. In: Thin Solid Films. European Materials Research Society (E-MRS) Spring Meeting 2013 Symposium O: Synthesis, Processing and Characterization of Nanoscale Multi Functional Oxide Films IV, vol. 553, pp. 132–137 (2014). ISSN:0040-6090
7. Haglund, Jr., R.F., Lawrie, B.J., Mu, R.: Coupling of photoluminescent centers in ZnO to localized and propagating surface plasmons. In: Thin Solid Films. Proceedings of the EMRS 2009 Spring Meeting Symposium H: Synthesis, Processing and Characterization of Nanoscale Multi Functional Oxide Films II, vol. 518, pp. 4637–4643 (2010). ISSN:0040-6090
8. Heitmann, D.: Radiative decay of surface plasmons excited by fast electrons on periodically modulated silver surfaces. J. Phys. C: Solid State Phys. **10**, 397 (1977). ISSN:0022-3719
9. Hofmann, C.E., et al.: Plasmonic modes of annular nanoresonators imaged by spectrally resolved cathodoluminescence. Nano Lett. **7**, 3612–3617 (2007). ISSN:1530-6984

10. García de Abajo, F.J.: Optical excitations in electron microscopy. Rev. Mod. Phys. **82**, 209–275 (2010)
11. Losquin, A., Kociak, M.: Link between cathodoluminescence and electron energy loss spectroscopy and the radiative and full electromagnetic local density of states. ACS Photon. **2**, 1619–1627 (2015)
12. Chu, M.-W., et al.: Probing bright and dark surface-plasmon modes in individual and coupled noble metal nanoparticles using an electron beam. Nano Lett. **9**, 399–404 (2009). ISSN:1530-6984
13. Mazzucco, S., et al.: Ultralocal modification of surface plasmons properties in silver nanocubes. Nano Lett. **12**, 1288–1294 (2012). ISSN:1530-6984
14. Myroshnychenko, V., et al.: Plasmon spectroscopy and imaging of individual gold nanodecahedra: a combined optical microscopy, cathodoluminescence, and electron energy-loss spectroscopy study. Nano Lett. **12**, 4172–4180 (2012). ISSN:1530-6984
15. Koh, A.L., et al.: Electron energy-loss spectroscopy (EELS) of surface plasmons in single silver nanoparticles and dimers: influence of beam damage and mapping of dark modes. ACS Nano **3**, 3015–3022 (2009). ISSN:1936-0851
16. Zhu, X.L., et al.: Confined three-dimensional plasmon modes inside a ring-shaped nanocavity on a silver film imaged by cathodoluminescence microscopy. Phys. Rev. Lett. **105**, 127402 (2010)
17. Coenen, T., Arango, F.B., Koenderink, A.F., Polman, A.: Directional emission from a single plasmonic scatterer. Nat. Commun. **5** (2014). https://doi.org/10.1038/ncomms4250. http://www.nature.com.proxy.library.vanderbilt.edu/ncomms/2014/140203/ncomms4250/full/ncomms4250.html
18. Bosman, M., Keast, V.J., Watanabe, M., Maaroof, A.I., Cortie, M.B.: Mapping surface plasmons at the nanometre scale with an electron beam. Nanotechnology **18**, 165505 (2007). ISSN:0957-4484, 1361-6528
19. Guiton, B.S., et al.: Correlated optical measurements and plasmon mapping of silver nanorods. Nano Lett. **11**, 3482–3488 (2011). ISSN:1530-6984
20. Li, G., et al.: Examining substrate-induced plasmon mode splitting and localization in truncated silver nanospheres with electron energy loss spectroscopy. J. Phys. Chem. Lett. **6**, 2569–2576 (2015). ISSN: 1948-7185
21. Coenen, T., Vesseur, E.J.R., Polman, A., Koenderink, A.F.: Directional emission from plasmonic Yagi-Uda antennas probed by angle-resolved cathodoluminescence spectroscopy. Nano Lett. **11**, 3779–3784 (2011). ISSN:1530-6984
22. Kim, J., Lee, G.J., Park, I., Lee, Y.P.: Finite-difference time-domain numerical simulation study on the optical properties of silver nanocomposites. J. Nanosci. Nanotechnol. **12**, 5527–5531 (2012)
23. Yang, Y.Y., Egerton, R.F.: Tests of two alternative methods for measuring specimen thickness in a transmission electron microscope. Micron **26**, 1–5 (1995). ISSN:0968-4328
24. Iakoubovskii, K., Mitsuishi, K., Nakayama, Y., Furuya, K.: Mean free path of inelastic electron scattering in elemental solids and oxides using transmission electron microscopy: atomic number dependent oscillatory behavior. Phys. Rev. B **77**, 104102 (2008)
25. Malis, T., Cheng, S.C., Egerton, R.F.: EELS log-ratio technique for specimen thickness measurement in the TEM. J. Electron Microsc. Tech. **8**, 193–200 (1988). ISSN:1553-0817
26. Egerton, R.F.: Electron Energy-Loss Spectroscopy in the Electron Microscope, 3rd edn. 491 pp. Springer, New York (2011). ISBN:1-4419-9582-X
27. Losquin, A., et al.: Experimental evidence of nanometer-scale confinement of plasmonic eigenmodes responsible for hot spots in random metallic films. Phys. Rev. B **88**, 115427 (2013)
28. Bosman, M., et al.: Encapsulated annealing: enhancing the plasmon quality factor in lithographically–defined nanostructures. Sci. Rep. **4** (2014/2015), ISSN:2045-2322. https://doi.org/10.1038/srep05537. http://www.nature.com/articles/srep05537
29. Talebi, N., et al.: Excitation of mesoscopic plasmonic tapers by relativistic electrons: phase matching versus eigenmode resonances. ACS Nano **9**, 7641–7648 (2015). ISSN:1936-0851
30. McMahon, M.D., Lopez, R., Meyer III, H.M., Feldman, L.C., Haglund, Jr., R.F.: Rapid tarnishing of silver nanoparticles in ambient laboratory air. Appl. Phys. B **80**, 915–921 (2005). ISSN:0946-2171, 1432-0649

Chapter 6
The Plasmonic Response of Archimedean Spirals

As previously discussed in Sect. 1.1, FIB- and lithography-based synthesis techniques have been utilized to assemble complex plasmonic nanostructures, with exceptional control of the nanoscale optical properties [1–8]. A plasmonic nanostructure of recent interest is the Archimedean nanospiral. The appeal of the structure is based off of its demonstrated ability to sustain resonant modes with high polarization dependencies over a wide range of wavelengths [9]. Beyond the linear response, the nanospiral has been shown to support strong second-harmonic generation (SHG) [10], produce super-continuum plasmonic emission [11], and generate orbital angular momentum states [12].

However, observing the near-field response of plasmonic nanostructures is critical to the optimization of the nanostructure geometry. The properties of the nanospiral have largely been explored using FDTD and far-field optical experiments [13–16]. However, recently techniques such as scanning near-field optical microscopy [17, 18] and CL [19, 20] have been used to observe the unique near-field plasmonic response of complex nanostructures.

In this chapter, I utilize STEM-CL to observe the plasmonic Archimedean spiral structures from two perspectives: metal spirals deposited on an electron-transparent substrate (nanospirals) and spiral slits milled into a metal substrate (spiral holes). For the nanospirals, I examine different plasmon modes present and demonstrate how STEM-CL can be combined with photonics techniques to produce high-resolution maps of the near-field modes. For the spiral holes, STEM-CL is used to observe orbital angular momentum (OAM) modes that are generated through interfering SPPs. For both types of structures, it is demonstrated that utilizing beam-induced light as opposed to beam electrons provides some significant advantages for plasmonic analyses.

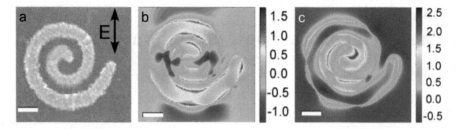

Fig. 6.1 Unique plasmon modes in archimedean nanospirals. (**a**) SEM image of a plasmonic Archimedean nanospiral. (Reprinted with permission from [9]. Copyright 2010 American Chemical Society.) (**b**) FDTD simulation of the E-field enhancement at 601 nm (visible regime) shows a plasmon mode called the "hourglass" mode. (**c**) FDTD simulation of the enhancement at 875 nm in the NIR shows a plasmon mode called the "focusing" mode. The spatial profile of the plasmonic response highly varies with respect to the spectral regime. Scale bar is 100 nm

6.1 Combining Photonics and Electron Microscopy for Plasmonic Analyses

The complex geometry of the nanospiral makes it a desirable structure for plasmonic analysis for a couple of different reasons. Firstly, the structure is highly tunable, as there are a wide variety of parameters such as arm width and spacing, winding number, thickness, and chirality that significantly affect the optical response. Additionally, the nanospiral exhibits the plasmonic response of both a single nanostructure and multiple interacting nanostructures simultaneously, through the surface charge either oscillating along the spiral arms or between them. The tunability and geometry allow for the generation of a wide range of plasmonic mode that cover a broad spectral range and possess unique spatial characteristics [9].

Figure 6.1 contains an SEM image of a nanospiral (a) and FDTD (b and c) simulations of the plasmonic near-field distribution, both of which were presented by Dr. Jed Ziegler in [9]. The FDTD simulations are the response to a plane wave polarized in the y-direction (as indicated by the arrow in Fig. 6.1a). Figure 6.1b shows the E-field enhancement at a wavelength of 601 nm; this plasmon mode is termed the "hourglass" mode and is predominantly active in the visible regime. Figure 6.1c shows the E-field enhancement at 875 nm. This mode is called the "focusing" mode and is usually found in the near infrared (NIR). The two plasmon modes are just a portion of the total optical response of the nanospirals (a more complete description can be found in [14]), but they are good examples of plasmon modes with highly varying spatial character across different wavelength regimes in the same sample. Additionally, the hourglass mode can have its plasmonic response replicated by a series of ellipses with the same spacing as the arms [9], while the focusing mode can only be produced in a continuous structure. Thus, these modes can be used to represent the single-structure/multiple-structure duality of the nanospiral and are ideal for demonstrating the versatility of the plasmonic response of the nanospiral.

Modeling the optical response through simulations provides an excellent starting point, but nanoscale investigations are critical for confirming the theoretical models as well as providing nanoscale control over specific excitations. Toward this end, I fabricate the nanospirals through an electron-beam lithography (EBL)-based process that generates Si-substrate chips with a thin SiN window, with the nanostructures patterned directly on the surface. The fabrication process was originally developed by Roderick Davidson with the help from Dr. Scott Retterer (an outline of the process is provided in Appendix C.2).

6.1.1 EELS Analysis of Lithographically Prepared Nanostructures

First, the nanospiral is examined through EELS. For these experiments, I used the Zeiss Libra200MC monochromated TEM/STEM at the University of Tennessee at Knoxville operated at an accelerating voltage of 200 keV, with a FWHM on the ZLP of 170 meV. The sample is made with 80-nm-thick Au, with 40-nm-wide arms, 60 nm separation between arms, and a winding number of 4π. Figure 6.2a shows a HAADF image of the structure examined. The EELS analysis shows peaks at ~1.6 eV (774 nm – NIR) and ~2.4 eV (516 nm – vis), and the SI of the two peaks are shown in Fig. 6.2b, c, respectively.

The 1.6-eV EELS peak matches the spatial profile of the FDTD simulation of the focusing mode shown in Fig. 6.1b relatively well. Furthermore, the peak is found in the NIR where the focusing mode is predicted to be supported. The spatial and spectral match of the EELS and FDTD indicates that the SI in Fig. 6.2b is an experimental map of the focusing plasmon mode.

However, while the EELS peak at ~2.4-eV EELS peak matches the hourglass mode spectrally, it does not match the spatial profile of the FDTD simulation from Fig. 6.1c. To better understand the disparity between the FDTD simulations and the EEL-SI, it is helpful to examine the individual spectra.

The SI in Fig. 6.2b, c each have ROIs marked with colored boxes; here I take the spectra from those ROIs and fit the peaks with Lorentzians and power laws for the backgrounds. The spectra within the red box in Fig. 6.2b (from the inner tip of the nanospiral) are plotted in Fig. 6.2d. The EELS peak corresponding to the plasmon mode is clearly visible. Furthermore, the peak is isolated spectrally from other low-loss features such that power-law background subtraction accurately captures the behavior of the peak, and the spatial profile of the plasmon mode can be observed in the SI.

The same cannot be said for the 2.4-eV peak. Figure 6.2e, f shows the spectra (and fits) for the ROIs in yellow and magenta shown at the top of Fig. 6.2c. It can be seen that the EELS peak shifts between pixels in this region, and considering that Au is shown to have bulk losses from interband transitions in this spectral regime, it is difficult to attribute the detected signal in the SI to the genuine behavior of LSPR modes here [21, 22].

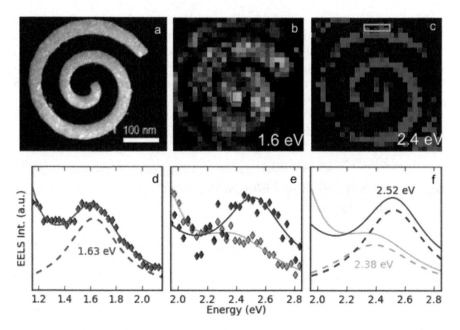

Fig. 6.2 Analyzing the plasmon response with EELS. (**a**) A HAADF image of an Au Archimedean nanospiral with 40-nm-wide arms, 60 nm arm spacing, 4π winding, and 80 nm thickness. (**b**) and (**c**) show the SI maps for plasmon peaks at (**b**) 1.6 eV and (**c**) 2.4 eV. (**d**) shows the spectra from the red ROI in (**b**) showing that the focusing mode has a strong peak. While (**e**) and (**f**) show the spectra (**e**) and the fits (**f**) from the yellow and magenta ROIs in (**c**), showing that the EELS signal in this region is not coming from a single feature

More importantly there is a fundamental barrier preventing the EEL-SI from duplicating the plasmon response demonstrated in the FDTD simulations, which is that the hourglass mode has a strong linear-polarization response [14]. Since, the electron beam is not polarized, but can excite polarization-dependent modes, the signal convolves the response of all of the polarization modes simultaneously.

6.1.2 Enhancing STEM with Photonics

STEM-CL can also be used to access plasmon modes at the near field but has its own set of drawbacks. The two principal modes of STEM-CL acquisition are via a spectrometer (CCD paired with a diffraction grating) or a PMT. For spectroscopy, both the CCD and the diffraction grating have their collection efficiencies fall-off across a moderate bandwidth, and additionally the pixelated CCDs are fairly vulnerable to electronics noise. PMTs, on the other hand, provide modestly higher-collection efficiencies and reduce the electronics noise significantly by having only a single pixel, making it much more effective in detecting weak signals.

Unfortunately, the PMT does not have any spectral sensitivity, so the light across the entire detectable spectral range is integrated into the signal. So both modes of CL acquisition present some inherent issues for detecting and mapping individual plasmon modes in nanostructures with weaker signals.

However, there are a few aspects of STEM-CL that provide some advantages with respect to EELS for low-signal analysis. By using the light emitted from the plasmon excitations as opposed to the EEL in the beam electrons, it allows for the radiative features to be analyzed directly without being influenced by the signal from non-radiative effects. Secondly, the light-based signals allow for the combination of photonics and electron microscopy to achieve the spectral resolution of the former with the spatial resolution of the latter. By combining the low electronics noise of the PMT with polarization and spectral filtering, one enables the isolation of and high-efficiency mapping of individual plasmon modes.

While EELS was able to reproduce the response of the focusing mode predicted through FDTD, the result can be improved through the spectrally filtered CL. A VG-HB601 STEM equipped with a parabolic mirror for CL collection, operating at 60 kV, is used for the experiments, with a high beam current (~2 nA) to increase the detected signal. Figure 6.3a, b shows the DF image and the PMT-CL image of a nanospiral with the same dimensions as in Sect. 6.1.1.

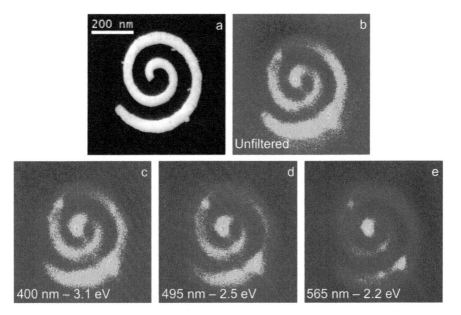

Fig. 6.3 Long-pass filtered CL. (**a**) A DF image of a nanospiral with the same dimensions as Fig. 6.2 and (**b**) the unfiltered PMT-CL image of the nanospiral. To access the focusing mode, the visible emission must be spectrally filtered. (**c**)–(**e**) show the different long-pass filtered images for (**c**) 400 nm (3.1 eV), (**d**) 495 nm (2.5 eV), and (**e**) 565 nm (2.2 eV). Once the visible emission has been filtered out, the bright spot at the inner tip remains, and a high-resolution map of the focusing mode is shown

The PMT-CL image in Fig. 6.3b shows the emission intensity across the whole spectrum. The fact that the peak intensity seems to originate from the center indicates that the focusing mode is strongly active in the sample, but without spectral resolution, it cannot be confirmed that the focusing mode is the cause for the luminescence there. In order to remove the extraneous signal, the CL output is filtered with long-pass filters that block out any signal with shorter wavelengths than the cutoff value.

Figure 6.3c–e shows three different long-pass-filtered, PMT-CL images for three different cutoff frequencies: 400 nm – 3.1 eV (c), 495 nm – 2.5 eV (d), and 565 nm – 2.2 eV (e). From Sect. 6.1.1, recall that peaks in the visible signal were present only at 2.3 eV and above, meaning that for the 565 nm long-pass filter, all of these modes are removed. Surely enough, without the signal from the peaks in the visible regime, only the high intensity at the tip of the nanospiral remains, indicating that the focusing mode is now the dominant feature in the PMT-CL image. Figure 6.3e is a 256×256 pixel image, while the SI in Fig. 6.2 was a 27×27 pixel image resulting in a much higher-resolution image. Furthermore, the PMT-CL image is acquired in three and a half minutes, while the SI is acquired in 12 min and 10 s, demonstrating the efficient acquisition of spectrally filtered CL.

Additionally, the usage of the radiative emission of optical excitations allows us to capture polarization effects with nanoscale precision, allowing us to access the hourglass mode that was missing in the EELS analysis. Here, instead of spectral filters, a linear polarizer is used and rotated with respect to the stationary sample.

Figure 6.4a shows an Au nanospiral with 40-nm-wide arms, 60 nm arm spacing, and a reduced thickness of 40 nm to cut down on bulk effects and focus only on the plasmonic response related to the spiral geometry. The PMT-CL image of the nanospiral shown in Fig. 6.4b exhibits a reduced signal-to-noise ratio compared with that seen in Fig. 6.3 as a result of the reduced thickness. However, now the CL signal is localized along the edges of the arm, just as it was in Fig. 6.1, a feature that was not observed in the thicker 80 nm samples.

The inclusion of a polarizer in the CL signal immediately shows the profile of the hourglass mode and produces a high-resolution map of the isolated plasmon (shown in Fig. 6.4c). Spectral filtering is not required for this mode, as it is the only optical feature with linear-polarization dependence, so all emission sources other than the hourglass mode aligned with the linear polarizer are attenuated by the filter [14]. By rotating the polarizer by 45° (Fig. 6.4d) and 90° (Fig. 6.4e), it can be seen that different polarizations of the hourglass mode can be easily isolated.

It is important to note that spectrally filtering this sample does not result in the observation of the focusing mode (like it did in Fig. 6.3), meaning that the focusing mode is not strongly active in this sample. The absence of the focusing mode is likely due to the strong deformity in the inner arm that can be seen in Fig. 6.4a. The discontinuity is a result of an undiagnosed error in the fabrication process that occurs when thin (and small) samples are deposited using the process outlined in Appendix C.2.

6.2 Orbital Angular Momentum in Plasmonic Spiral Holes

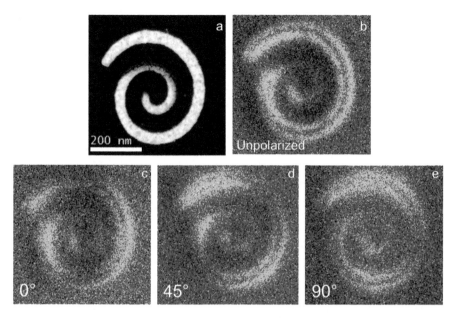

Fig. 6.4 Linearly-polarized CL. (**a**) A DF image of a nanospiral with the same lateral dimensions as the last two figures, but half the total thickness (40 nm). (**b**) the unpolarized PMT-CL image of the nanospiral shows that the signal-to-noise has decreased, but the localization of the plasmon mode to the spiral edges, as seen in Fig. 6.1 has appeared. By applying a linear polarizer (**c**) and rotating it to 45° (**d**) and 90° (**e**), different polarizations of the hourglass mode can be accessed

6.2 Orbital Angular Momentum in Plasmonic Spiral Holes

In this section, I employ STEM-CL, combined with FDTD simulations, to demonstrate the near-field interaction between OAM modes in plasmonic nanostructures and chiral substructures. The structures in this section were all prepared by Dr. Benjamin Lawrie and were fabricated by depositing 100 nm film of Ag on a 25 nm SiN membrane identical to those in Sect. 6.1. A FIB is then used to mill a thin hole in the Ag film (that does not penetrate the SiN to maintain the stability of the window) in the shape of an Archimedean nanospiral. The geometry is known to generate optical OAM modes, and the CL response of the structures is compared to FDTD simulations of the LDOS to confirm the presence of OAM. Finally, the interaction between chiral nanostructures and OAM modes is explored by milling nanospirals, an order-of-magnitude smaller in size, at the center of the larger spiral. The luminescence of the composite system is found to be significantly higher in systems where the substructure has the same chirality as the outer structure, providing spatially and spectrally resolved evidence of coupling between OAM and chiral nanostructures.

I took interest in these structures because, light possessing OAM has been a topic of considerable interest in recent years [23–27]. This interest has been fueled by critical advances in optical manipulation and trapping [28, 29] that leverage the helical wavefront of optical OAM modes to impart angular momentum into micro- and nanoscale structures. Additionally, the added degrees of freedom provided by the orthonormal angular momentum basis set enable large-scale multiplexing of classical and quantum communications and provide a framework for emerging studies of quantum information science [30–34]. Furthermore, it has been proposed that OAM modes can be used for the detection of molecular chirality [35–37], which plays a dominant role in biological and chemical processes [38].

Surface plasmons are routinely used to manipulate light at the nanoscale, resulting in the development of a wide class of metasurfaces and asymmetric plasmonic nanostructures that generate and control OAM on chip [12, 39–44]. Techniques such as near-field scanning optical microscopy have been used to experimentally observe nanoscale OAM effects in plasmonic nanostructures [45–49]. For instance, Chen et al. have recently shown the ability to actively control OAM with polarization-specific near-field spectroscopy [18]. These recent developments illustrate the potential for near-field applications of OAM modes in nanotechnology.

In order to generate a OAM mode in the structure, the geometry of the spiral must be chosen precisely. The formula for an Archimedean spiral in polar coordinates is $r(\theta) = r_0 + d \cdot \theta$ where r_0 is the initial radius and d is the distance between the arms. An example of the geometry of a spiral hole is shown in Fig. 6.5a. The generation of OAM in plasmonic spiral holes comes when the phase of SPPs interfere coherently such that the composite plasmonic response can be described by a Bessel function whose topological charge, ℓ, is determined by the ratio, $\ell = 2\pi d/\lambda_{SPP}$, where λ_{SPP} is the surface plasmon polariton wavelength in the system. Each point on the surface of the slit serves as a point source for surface plasmon polaritons, and the polaritons are launched in all directions from the entire slit. If the dimensions are carefully chosen such that the arm spacing, d, is an integer multiple of λ_{SPP}, then there is a 2π offset between SPPs coming from the inner and outer arms, heading toward the origin of the spiral. Additionally, if the initial radius r_0 is an integer multiple of λ_{SPP}, then for every position along the inner edge of the slit, SPPs that propagate toward the origin all contribute to the phase singularity. Figure 6.5b–d show examples of spiral holes that should have OAM states with topological order, $m = 1$ (b), $m = 2$ (c), and $m = 3$ (d).

If both r_0 and d are equal to λ_{SPP}, then the phase mismatch is 2π and the order of the OAM state, m_{OAM}, is 1; an example is shown in Fig. 6.5b. However, the topological charge of the OAM can be increased by increasing the distance between the inner and outer arms and generating a larger phase shift. If $d = 2\lambda_{SPP}$ or $3\lambda_{SPP}$, then the charge of the OAM state can be increased to 2 or 3, respectively (Fig. 6.5b, c).

The generation of the OAM state can be modeled via FDTD simulations. Figure 6.6a shows a spiral with $r_0 = d = 660$ nm, which is determined to be λ_{SPP} for a three layer system (air, Ag, SiN) [50]. Figure 6.6b shows the FDTD simulation, performed by Prof. Sang-Yeon Cho, of the phase distribution for the E-field, when excited by a plane wave source polarized in the y-direction. At the origin of the spiral, the phase singularity and E-field amplitude null point characteristic of OAM modes are seen in Fig. 6.6b, c.

6.2 Orbital Angular Momentum in Plasmonic Spiral Holes

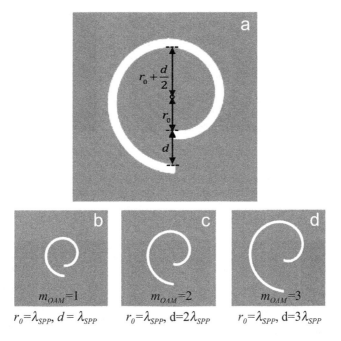

Fig. 6.5 Surface plasmon polaritons in a spiral hole. (**a**) An example of an Archimedean spiral, $r(\theta) = r_0 + d\frac{\theta}{2\pi}$. Orbital angular momentum (OAM) states can be generated by choosing r_0 and d to be integer multiples of the SPP wavelength, λ_{SPP}, generating a phase singularity at the origin of the spiral. By increasing the arm separation to higher integer multiples of the SPP wavelength, $d = n\lambda_{\text{SPP}}$, different orders of OAM, m_{OAM} can be achieved. (**b**) $n = 1, m_{\text{OAM}} = 1$, (**c**) $n = 2, m_{\text{OAM}} = 2$, (**d**) $n = 3, m_{\text{OAM}} = 3$

6.2.1 Visualizing Orbital Angular Momentum with Cathodoluminescence

To observe the effects at an experimental level, holes with arm separation, $d = 1\lambda_{\text{SPP}}$, $d = 2\lambda_{\text{SPP}}$, and $d = 3\lambda_{\text{SPP}}$, are prepared and examined in a VG-HB601 equipped with a parabolic mirror for CL, and operating at 60 kV, with a beam current of roughly 2 nA.

The plasmonic response of the different spiral holes are shown in Fig. 6.7. BF images of all the spiral holes are shown in Fig. 6.7a–c. The remaining images in the figure are all band-pass-filtered PMT-CL images, taken with band-pass filters of increasing bandpass wavelength, λ_{BP}. Figure 6.7d–f show bandpass-filtered images with λ_{BP}=445 nm, Fig. 6.7g–i the same spirals with $\lambda_{\text{BP}} = 513$ nm, Fig. 6.7g–i with $\lambda_{\text{BP}} = 586$ nm, and Fig. 6.7g–i with $\lambda_{\text{BP}} = 685$ nm.

The most immediately noticeable aspect of the CL images is the spiral interference pattern, the period of which increases with increasing wavelengths. Similar interference patterns have been previously observed for linear plasmonic gratings

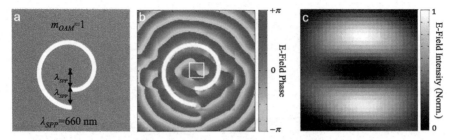

Fig. 6.6 Electric field phase and intensity for an OAM state. (a) A spiral with $r_0 = d = \lambda_{SPP} = 660$ nm, resulting in a $m_{OAM} = 1$ state. (b) The phase plot for the E-field, showing the phase singularity at the origin of the spiral. (c) A magnified plot of the normalized field intensity from the origin of the spiral, marked by the white box in (b). The trademark null point of the intensity profile shows that an OAM state is present

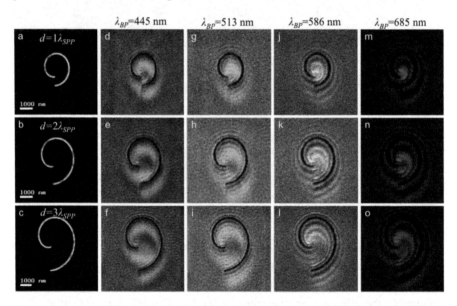

Fig. 6.7 Plasmonic response in spiral holes via cathodoluminescence. The CL response of λ, 2λ, and 3λ spiral holes. (a–c) BF images of the spiral holes. (d–o) Band-pass-filtered PMT-CL images of the 1λ, 2λ, and 3λ spirals with the band-pass wavelength, λ_{BP} at (d–f) 445 nm, (g–i) 513 nm, (j–l) 586 nm, and (m–o) 685 nm

and attributed to interference between SPPs and transition radiation (TR) caused by the electron impinging on the metal surface [51]. The TR is a result of a charged particle passing between two different dielectric media [52]. The transition radiation is azimuthally isotropic, while the SPP out-coupling is directionally dependent; the result is coherent interference between the two when detected at the far field. Kuttge et al. demonstrated this effect in linear gratings and were able to model it with the following equation [51]:

$$I_{CL} = \int_{mirror} d\Omega |A_{SPP}\mathbf{S}(\Omega)e^{i\phi} + \mathbf{f}_{TR}(\Omega)|^2 \tag{6.1}$$

where Ω is the angle of the emission, A_{SPP} is the SPP amplitude, $\mathbf{S}(\Omega)$ is the normalized in-plane far-field amplitude of the SPP after being scattered by the grating, $\mathbf{f}_{TR}$ is the far-field amplitude of the transition radiation, and ϕ is the phase difference between the SPP and transition radiation. The importance of Eq. (6.1) lies in the $e^{i\phi}$ term, which shows that information about the phase of the SPPs in the sample is carried in the observed interference pattern.

The phase behavior from the FDTD simulations in Fig. 6.6b is not reproduced because the fringes here are not due to the phase of the SPP, but rather the phase difference of the SPP and the TR; however, they still possess information about the phase of the SPPs. Hence the spiral-like interference patterns within the arms of the spiral show experimentally that SPPs are interfering constructively and destructively in this region.

Additionally, the CL intensity is more directly comparable to the E-field intensity. Recall from Fig. 6.6c that at the origin of the spiral, there should be a dark spot in terms of the intensity for the OAM. Figure 6.8 shows the band-pass-filtered PMT-CL image, with $\lambda_{BP} = 685$ nm, for spirals with $d = 1\lambda$, $d = 2\lambda$, and $d = 3\lambda$. Figure 6.8a–c show the CL images, each with a dashed line that marks a line profile of the normalized CL intensity shown in Fig. 6.8d–f. There are three annotations on each of the CL images. The lower two denote the edges of the inner and outer arms, but the third represents a dip in intensity, observable for all three structures, that is fairly close to the origin of the spiral where the OAM state is expected to be observed.

The line profile in Fig. 6.8d shows the spacing between the inner and outer arms, as well as the spacing between the inner arm and the dip in intensity at the origin. Both distances are ~660 nm, or λ_{SPP}. The null point is only faintly observable in the $1\lambda_{SPP}$ spiral; however, looking at both the $2\lambda_{SPP}$ (Fig. 6.8e) and the $3\lambda_{SPP}$ spiral (Fig. 6.8f), the dip is far more pronounced. Additionally, for all three structures, the null point occurs at a distance of λ_{SPP} away from the inner arm at the origin of the spiral.

The presence of the null point in all three spirals, at the origin where an OAM state would generate a dip in the intensity at the OAM wavelength, demonstrates that the spirals do indeed produce OAM plasmon modes and that they can be detected through band-pass-filtered CL.

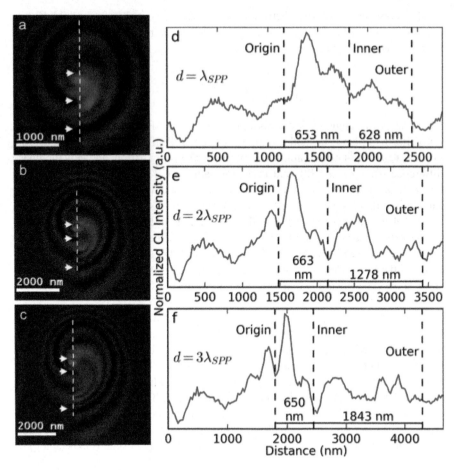

Fig. 6.8 Detecting orbital angular momentum with cathodoluminescence. (**a–c**) The band-pass-filtered PMT-CL images of the 1λ, 2λ, and 3λ spiral holes are shown for $\lambda_{BP} = 685$ nm (closest to λ_{SPP}). (**d–f**) Line profiles of the total CL intensity of the images shown in (**a–c**). All six subfigures have three features marked, the inner arm, the outer arm, and a dip in the CL intensity at the origin. The fact that all three spirals generate a dark spot at point almost exactly λ_{SPP} nm away from the inner arm shows that the spirals have successfully created OAM and that it is observable in CL

References

1. Ozbay, E.: Plasmonics: merging photonics and electronics at nanoscale dimensions. Science **311**, 189–193 (2006). ISSN: 0036-8075, 1095-9203
2. Kuttge, M., García de Abajo, F.J., Polman, A.: Ultrasmall mode volume plasmonic nanodisk resonators. Nano Lett. **10**, 1537–1541 (2009). ISSN: 1530-6984
3. Kuttge, M., Vesseur, E.J.R., Polman, A.: Fabry-Pérot resonators for surface plasmon polaritons probed by cathodoluminescence. Appl. Phys. Lett. **94**, 183104 (2009). ISSN: 0003-6951, 1077-3118

References

4. Bosman, M., et al.: Encapsulated annealing: enhancing the plasmon quality factor in lithographically–defined nanostructures. Sci. Rep. **4**, 5537 (2014/2015). ISSN: 2045-2322. https://doi.org/10.1038/srep05537. http://www.nature.com/articles/srep05537
5. Haberfehlner, G., et al.: Correlated 3D nanoscale mapping and simulation of coupled plasmonic nanoparticles. Nano Lett. **15**, 7726–7730 (2015). ISSN: 1530-6984
6. Chamuah, N., Nath, P.: Periodically varying height in metal nano-pillars for enhanced generation of localized surface plasmon field. Plasmonics **10**, 1367–1372 (2015)
7. Lau, U.Y., Saxer, S.S., Lee, J., Bat, E., Maynard, H.D.: Direct write protein patterns for multiplexed cytokine detection from live cells using electron beam lithography. ACS Nano **10**, 723–729 (2015)
8. Jonušauskas, L., et al.: Plasmon assisted 3D microstructuring of gold nanoparticledoped polymers. Nanotechnology **27**, 154001 (2016)
9. Ziegler, J.I., Haglund, R.F.: Plasmonic response of nanoscale spirals. Nano Lett. **10**, 3013–3018 (2010). ISSN: 1530-6984
10. Davidson, I.R.B., et al.: Efficient forward second-harmonic generation from planar archimedean nanospirals. Nanophotonics **4**, 108–113 (2015/2016). ISSN: 2192–8614. https://doi.org/10.1515/nanoph-2015-0002. http://www.degruyter.com/view/j/nanoph.2015.4.issue-1/nanoph-2015-0002/nanoph-2015-0002.xml?ncid=txtlnkusaolp00000603&format=INT
11. Krasavin, A., Ginzburg, P., Wurtz, G., Zayats, A.: Nonlocality-driven supercontinuum white light generation in plasmonic nanostructures. Nat. Commun. **7**, 11497 (2016)
12. Ohno, T., Miyanishi, S.: Study of surface plasmon chirality induced by Archimedes' spiral grooves. Opt. Express **14**, 6285–6290 (2006)
13. Tzuang, L.D.-C., et al.: Polarization rotation of shape resonance in Archimedean spiral slots. Appl. Phys. Lett. **94**, 091912 (2009)
14. Ziegler, J.I., Haglund, R.F., Jr.: Complex polarization response in plasmonic nanospirals. Plasmonics **8**, 571–579 (2012). ISSN: 1557–1955, 1557–1963
15. Ku, C.-D., Huang, W.-L., Huang, J.-S., Huang, C.-B.: Deterministic synthesis of optical vortices in tailored plasmonic archimedes spiral. IEEE Photonics J. **5**, 4800409–4800409 (2013)
16. Tsai, W.-Y., Huang, J.-S., Huang, C.-B.: Selective trapping or rotation of isotropic dielectric microparticles by optical near field in a plasmonic archimedes spiral. Nano Lett. **14** (2014), 547–552
17. Chen, W., Abeysinghe, D.C., Nelson, R.L., Zhan, Q.: Experimental confirmation of miniature spiral plasmonic lens as a circular polarization analyzer. Nano Lett. **10**, 2075–2079 (2010)
18. Chen, C.-F., et al.: Creating optical near-field orbital angular momentum in a gold metasurface. Nano Lett. **15**, 2746–2750 (2015)
19. Osorio, C.I., Coenen, T., Brenny, B.J., Polman, A., Koenderink, A.F.: Angleresolved cathodoluminescence imaging polarimetry. ACS Photonics **3**, 147–154 (2015)
20. Fang, Y., Verre, R., Shao, L., Nordlander, P., Kall, M.: Hot electron generation and cathodoluminescence nanoscopy of chiral split ring resonators. Nano Lett. **16**, 5183–5190 (2016)
21. Bosman, M., Keast, V.J., Watanabe, M., Maaroof, A.I., Cortie, M.B.: Mapping surface plasmons at the nanometre scale with an electron beam. Nanotechnology **18**, 165505 (2007). ISSN: 0957-4484, 1361-6528
22. Alkauskas, A., Schneider, S.D., Hébert, C., Sagmeister, S., Draxl, C.: Dynamic structure factors of Cu, Ag, and Au: comparative study from first principles. Phys. Rev. B **88** (2013), 195124
23. Marrucci, L., Manzo, C., Paparo, D.: Optical spin-to-orbital angular momentum conversion in inhomogeneous anisotropic media. Phys. Rev. Lett. **96**, 163905 (2006)
24. Yu, N., et al.: Light propagation with phase discontinuities: generalized laws of reflection and refraction. Science **334**, 333–337 (2011)
25. Cai, X., et al.: Integrated compact optical vortex beam emitters. Science **338**, 363–366 (2012)
26. Naidoo, D., et al.: Controlled generation of higher-order Poincaré sphere beams from a laser. Nat. Photonics **10**, 327–332 (2016)
27. Hachtel, J.A., et al.: Spatially and spectrally resolved orbital angular momentum interactions in plasmonic vortex generators. arXiv preprint. arXiv:1705.10640 (2017)

28. Andersen, M., et al.: Quantized rotation of atoms from photons with orbital angular momentum. Phys. Rev. Lett. **97**, 170406 (2006)
29. Padgett, M., Bowman, R.: Tweezers with a twist. Nat. Photonics **5**, 343–348 (2011)
30. Paterson, C.: Atmospheric turbulence and orbital angular momentum of single photons for optical communication. Phys. Rev. Lett. **94**, 153901 (2005)
31. Marino, A., et al.: Delocalized correlations in twin light beams with orbital angular momentum. Phys. Rev. Lett. **101**, 093602 (2008)
32. Wang, J., et al.: Terabit free-space data transmission employing orbital angular momentum multiplexing. Nat. Photonics **6**, 488–496 (2012)
33. Tamburini, F., et al.: Encoding many channels on the same frequency through radio vorticity: first experimental test. New J. Phys. **14**, 033001 (2012)
34. Bozinovic, N., et al.: Terabit-scale orbital angular momentum mode division multiplexing in fibers. Science **340**, 1545–1548 (2013)
35. Alexandrescu, A., Cojoc, D., Di Fabrizio, E.: Mechanism of angular momentum exchange between molecules and Laguerre-Gaussian beams. Phys. Rev. Lett. **96**, 243001 (2006)
36. Mondal, P.K., Deb, B., Majumder, S.: Angular momentum transfer in interaction of Laguerre-Gaussian beams with atoms and molecules. Phys. Rev. A **89**, 063418 (2014)
37. Wu, T., Wang, R., Zhang, X.: Plasmon-induced strong interaction between chiral molecules and orbital angular momentum of light. Sci. Rep. **5**, 18003 (2015)
38. Patterson, D., Schnell, M., Doyle, J.M.: Enantiomer-specific detection of chiral molecules via microwave spectroscopy. Nature **497**, 475–477 (2013)
39. Kang, M., Chen, J., Wang, X.-L., Wang, H.-T.: Twisted vector field from an inhomogeneous and anisotropic metamaterial. JOSA B **29**, 572–576 (2012)
40. Huang, L., et al.: Dispersionless phase discontinuities for controlling light propagation. Nano Lett. **12**, 5750–5755 (2012)
41. Yin, X., Ye, Z., Rho, J., Wang, Y., Zhang, X.: Photonic spin Hall effect at metasurfaces. Science **339**, 1405–1407 (2013)
42. Gorodetski, Y., Drezet, A., Genet, C., Ebbesen, T.W.: Generating far-field orbital angular momenta from near-field optical chirality. Phys. Rev. Lett. **110**, 203906 (2013)
43. Maguid, E., et al.: Photonic spin-controlled multifunctional shared-aperture antenna array. Science **352**, 1202–1206 (2016)
44. Garoli, D., Zilio, P., Gorodetski, Y., Tantussi, F., De Angelis, F.: Optical vortex beam generator at nanoscale level. Sci. Rep. **6**, 29547 (2016)
45. Kim, H., et al.: Synthesis and dynamic switching of surface plasmon vortices with plasmonic vortex lens. Nano Lett. **10**, 529–536 (2010)
46. Shen, Z., et al.: Visualizing orbital angular momentum of plasmonic vortices. Opt. Lett. **37**, 4627–4629 (2012)
47. Liu, A.-P., et al.: Detecting orbital angular momentum through division-of-amplitude interference with a circular plasmonic lens. Sci. Rep. **3**, 2402 (2013)
48. Carli, M., Zilio, P., Garoli, D., Giorgis, V., Romanato, F.: Sub-wavelength confinement of the orbital angular momentum of light probed by plasmonic nanorods resonances. Opt. Express **22**, 26302–26311 (2014)
49. Garoli, D., Zilio, P., Gorodetski, Y., Tantussi, F., De Angelis, F.: Beaming of helical light from plasmonic vortices via adiabatically tapered nanotip. Nano Lett. **16**, 6636–6643 (2016)
50. Maier, S.A.: Plasmonics: Fundamentals and Applications. Springer Science & Business Media, Berlin (2007)
51. Kuttge, M., et al.: Local density of states, spectrum, and far-field interference of surface plasmon polaritons probed by cathodoluminescence. Phys. Rev. B **79**, 113405 (2009)
52. Yamamoto, N., Sugiyama, H., Toda, A.: Cherenkov and transition radiation from thin plate crystals detected in the transmission electron microscope. Proc. R. Soc. Lond. A Math. Phys. Eng. Sci. **452**, 2279–2301 (1996)

Chapter 7
Future Directions and Conclusion

7.1 Advanced Experiments for Nanoscale Optical Analyses

For the future directions of my work, I hope to continue my pursuits of understanding optical properties at the nanoscale and take both EELS and CL to the next level.

For EELS, the next level is monochromation. While, I utilized a monochromated Zeiss Libra during the course of my Ph.D., this instrument is limited to around 110 meV energy resolution, and 150–200 meV is more typical. By utilizing the newly developed monochromators from Nion in an UltraSTEM, energy resolutions of up to 8 meV have been achieved [1].

Figure 7.1a shows a HAADF image of a nanospiral, taken on a monochromated UltraSTEM at Nion Company in Kirkland, WA. Figure 7.1b shows an EEL spectrum taken from the bottom of the nanospiral. The microscope was operated at 60 keV, and with the monochromation the FWHM of the ZLP was 32 meV, and high-quality factor plasmon peaks (67 meV FWHM) were observed at an energy of 0.357 eV (corresponding to a wavelength of 3470 nm). The capability to see phenomena that far into the infrared opens up a brand-new world of possibilities in terms of infrared plasmonics and beyond.

In terms of CL, the future steps are to incorporate more advanced optical techniques utilizing the high-spatial resolution CL signal. Figure 7.2a shows the schematic of one such technique, called a second-order autocorrelation experiment. In this experiment, the CL signal is passed through a 50/50 beam splitter, and the two beams are passed to single-photon counting PMTs. By putting a varying delay on one detector with respect to the other, it is possible to find the time where the time-of-flight between the two detectors is equivalent.

The power of this method is it provides time-resolution for the inherently time-integrated technique of CL, because at the zero-delay point, the photon statistics of the optical excitations can be measured by looking for coincidences of photons reaching the two detectors. Preliminary experiments at ORNL are already being

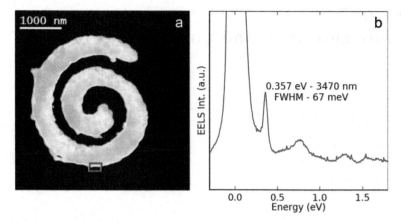

Fig. 7.1 Infrared plasmonics in the electron microscope. Breakthroughs in electron-beam monochromation from Nion Company allow for extremely high-energy resolution in EELS. (**a**) HAADF image of a nanospiral, (**b**) EEL spectrum from bottom of the nanospiral, showing an energy resolution of 32 meV, and the ability to detect plasmons deep into the infrared (0.357 eV to 3470 nm)

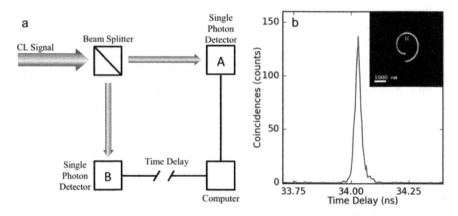

Fig. 7.2 Time-resolution in the electron microscope. (**a**) Schematic of an autocorrelation experiment, designed to study the photon statistics of optical excitations. (**b**) Correlation measurements from plasmonic excitations in a spiral hole (shown in inset). The plasmon shows photon bunching statistics, which could be due to multiple plasmon excitations or the electron source emitting with thermal statistics

performed Roderick B. Davidson II, Dr. Benjamin Lawrie, Dr. Raphael Pooser, and I, and we have successfully collected correlation data on plasmons from the spiral holes in Chap. 6, shown in Fig. 7.2b (BF image in inset).

There is a sharp peak visible at a time delay of ~34.1 ps that is a signal of photon bunching [2, 3]. The peak could be due to a number of different effects including electron bunching from the tip in the electron gun and multiple plasmons being excited with a single photon, and more research is required to understand these effects rigorously. However, the ability to study optical phenomena with high time- *and* spatial resolution presents many exciting opportunities for the future.

7.2 Outlook and Conclusion

Nanoscale effects and structures are at the heart of modern technology, and optical phenomena at the nanoscale comprise an important subset of the field. As synthesis and fabrication techniques become more and more advanced, the complexity of the fabricated devices becomes higher and higher, and needed spatial resolution for effective analysis becomes finer and finer. This has made techniques to access the effects of nanoscale variations highly important across technology as a whole.

The advent and advancement of nanoscale analysis tools, both theoretical and experimental, have spurred technology to even greater heights. Theoretical techniques, such as DFT and FDTD, have been around for decades, but with the highly developed commercial programs such as VASP and Lumerical Solutions Inc., the basic techniques have been brought to new levels. Similarly from the experimental side, one decade ago nobody had been able to map plasmon resonances with at the level of spatial resolution that electron microscopy now routinely achieves, and new advances in monochromation and specialty sample holders allow for access to even more elusive nanoscale physical effects.

The analysis of nanoscale optical phenomena, such as plasmons, *directly at the nanoscale*, is an especially important area of study, because of the role it plays within the community. Optics as a whole is a massive field of knowledge, but one where the predominant experimental techniques cannot access nanoscale effects. Super-resolution optical microscopy receiving the Nobel Prize in 2014 shows the commitment of the field to direct observation at the nanoscale and the importance that such measurements have to the world as a whole. The access to data hidden beneath the diffraction limit and complementarity to purely optical research ensures that the type of nanoscale analysis techniques demonstrated in this dissertation will stay at the forefront of the research community as the world moves forward to a more complex future.

References

1. Krivanek, O.L., et al.: Vibrational spectroscopy in the electron microscope. Nature **514**, 209–212 (2014)
2. Jeltes, T., et al.: Comparison of the Hanbury Brown-Twiss effect for bosons and fermions. Nature **445**, 402–405 (2007)
3. Meuret, S., et al.: Photon bunching in cathodoluminescence. Phys. Rev. Lett. **114**, 197401 (2015)

Appendix A
Overview of Electron Microscopes

Over the course of my Ph.D., I used many electron microscopes; in this appendix, however, I will discuss the three microscopes that produced all the data presented in this dissertation.

A.1 Nion UltraSTEM 200

The US200 is the workhorse machine of the STEM group at ORNL. It is a dedicated cold field emission gun (FEG) STEM with a fifth-order aberration corrector which allows for extremely high spatial resolution (~55 pm). The microscope also has a Gatan Enfinium Dual EELS spectrometer, which allows for EELS simultaneous acquisition in two energy ranges.

The microscope is extremely versatile because of how powerful it is. It is ideally suited for compositional analysis between the ability to perform high-resolution HAADF imaging, as well as doing quantitative EELS by being able to simultaneously acquire the low-loss and core-loss regimes. The cold FEG has an energy spread of ~350 meV at best, which allows for plasmonic analyses in the low-loss regime as well. Although for higher energy resolution, sometimes the monochromated Libra discussed in Sect. A.3 is used.

Most of imaging and EELS reported in this dissertation was performed on the US200. The example images and EELS in Sects. 2.3.2 and 2.4.2, and all microscopy results in Sects. 4.1–4.2.2, as well as all of the EELS and HAADF images in Chap. 5 were produced with this microscope.

A.2 VG-HB601

The 601 is a dedicated aberration-corrected STEM that has been specifically modified to perform CL. The general setup is similar to the US200; however, several parameters have been altered to improve the CL. The most important modification is the inclusion of a parabolic mirror below the sample. The mirror covers a 2 Str solid angle, with a small hole drilled through the top (to allow the electron beam to pass through). In order to optimize the mirror position, it is controlled in all three dimensions with piezoelectric motors. The mirror directs any emitted light out of a small port in the side of the microscope.

The other important aspect is that several important parameters have been changed in the microscope to maximize beam current (and hence CL signal) as opposed to spatial resolution. By changing the strengths of various lenses, I commonly change the beam current between 1 and 3 nA, compared to the ~25 pA in the US200. The cost of the increased signal is an increased probe size and worse spatial resolution. Even though the microscope is equipped with a third-order aberration corrector, the microscope cannot achieve atomic spatial resolution under the CL setup. The 601 has an EELS detector, but it is not used as the quality of EELS on the other microscopes is significantly higher.

The 601 is the only microscope capable of CL that I had access to, so all CL results shown in this dissertation come from the 601 along with the images in Sects. 6.1.2 and 6.2 and the coincidence measurements in Sect. 7.1.

A.3 Zeiss Libra200-MC

The Libra is different from the 601 and the US200 in two important ways. Firstly, the electron gun is a Schottky FEG instead of a cold FEG. While the energy spread and beam coherence of the cold FEG is stronger than a Schottky FEG, the stability of the beam current is stronger in the Schottky FEG as well having less noise in the signal. The higher energy spread is accommodated by the second main difference in the microscope: monochromation. The Libra is equipped with an electrostatic omega-type monochromator. The monochromator can reduce the ZLP FWHM of a Schottky FEG from 4–500 meV to 43 meV[1], but the monochromator possesses many different slits, so the balance between beam current and energy resolution can be optimized to the users' particular needs.

There are other significant differences as well. Firstly, the Libra is a TEM primarily that can be run in STEM mode; however, for my experiments, it was only operated in TEM mode for alignment procedures; all data was acquired in STEM mode. Secondly, the Libra is equipped with a MANDOLINE filter for energy-filtered EELS imaging; this is a powerful capacity, but not one that I used for my experiments.

The Libra is not aberration-corrected, so it was not used for any high-resolution imaging purposes (where the US200 would be better suited). However I was regularly able to achieve 150 meV energy resolution while still maintaining significant beam current, so it was ideally suited for plasmonic EELS analysis. The EELS plasmon maps and corresponding HAADF images shown in Sects. 4.2.3 and 6.1.1 are performed on this microscope.

Reference

1. Essers, E., Benner, G., Mandler, T., Meyer, S., Mittmann, D., Schnell, M., Höschen, R.: Energy resolution of an omega-type monochromator and imaging properties of the mandoline filter. Ultramicroscopy **110**(8), 971–980 (2010)

Appendix B
Fit Parameters EELS and CL Data in Chap. 5

In order to determine the peak positions in both EELS and CL, the raw data was fitted with the nonlinear least squares regression tool available in the SciPy Python Library, curvefit. All fit parameters for all spectra shown in the paper are included here in Table B.1. For all the CL peaks Lorentzians are fit to the data. It is important to note that the fitting is performed with respect to wavelength as opposed to energy, as that is the measured property in which the data are acquired; afterward the fit and the data are converted to energy in eV. For the EEL spectra, there is a strong background coming from the ZLP. While the tail of the ZLP is not precisely exponential, it can be fitted fairly accurately with a power-law background over the small energy ranges in which the fitting is performed (i.e., a few eV). Then, the remaining ZLP-tail-subtracted-EELS is fit with Lorentzians to determine peak positions. A total of five peaks are observed across both spectroscopies in the main text, all the parameters of which are collated in Table B.1.

There are a few important points to note in these values. Firstly, it can be seen in EELS that the weak plasmon peaks from the center of the nanoparticle and the SA are wider than the LA, transverse plasmons, and bulk plasmons. The explanation for this is likely the presence of other plasmons that are too weak to be resolved into individual peaks, but which broaden the EELS signal for the major plasmons nearby. Additionally, the amplitudes of the different fits are not necessarily directly comparable, as different numbers of pixels for the spectrum image have been summed for each spectrum plotted in the main text. For Fig. 5.1f, the plotted spectra contain 2 pixels from the CL-SI and 32 for the EELS-SI. For Fig. 5.1g, 3 CL pixels and 48 EELS pixels.

For Fig. 5.4d, both EEL spectra contain 9 pixels. And for Fig. 5.5d, the CL contains 1 CL-SI pixel and the EELS contains 9 EEL-SI pixels (the EEL spectrum in Fig. 5.4d Point 2 is taken from the same pixels as the spectrum in Fig. 5.5). It is also important to note that directly fitting a power-law background at Point 1 was not possible. As an energy range starting at approximately 1.4 eV was used to block out the majority of the ZLP intensity, to allow for longer acquisition times. At Point 1 the width of the 2.0-eV plasmon is large enough that there is not sufficient signal on the low energy side of the spectrum to accurately fit a power-law background. To account for the power law at Point 1, I use the same exponential value for the power-law background fitted at Point 2 and allow the fit to vary the amplitude and zero level. From the fits, it can be seen that this results in rather higher peak intensities, but the peak positions are believed to be correct, due to the strong fit of the data.

Table B.1 Parameters for CL and EELS fits

Parameters for Lorentzian peak fits			$f(\omega) = A \cdot \frac{\gamma}{(\omega-\omega_0)^2+\gamma^2}$		
Figure	Spectroscopy	Plasmon	ω_0	A	γ
Figure 5.1f	CL	LA	1.95 eV (634.44 nm)	1.34×10^4	60.89 nm
Figure 5.1f	CL	OOP	2.53 eV (489.69 nm)	1.49×10^4	111.9 nm
Figure 5.1g	CL	LA	1.92 eV (645.50 nm)	5.96×10^3	87.90 nm
Figure 5.1g	CL	OOP	2.45 eV (506.10 nm)	7.52×10^3	56.91 nm
Figure 5.1g	CL	SA	3.02 eV (410.53 nm)	3.28×10^3	40.60 nm
Figure 5.5d	CL	LA	1.92 eV (645.30 nm)	1.51×10^4	90.89 nm
Figure 5.5d	CL	OOP	2.53 eV (489.58 nm)	4.78×10^3	63.37 nm
Figure 5.1f	EELS	LA	2.03 eV	1.33×10^6	0.57 eV
Figure 5.1f	EELS	Trans	3.56 eV	3.21×10^6	1.05 eV
Figure 5.1g	EELS	SA	2.85 eV	1.32×10^6	0.83 eV
Figure 5.1g	EELS	Trans	3.57 eV	7.65×10^5	0.32 eV
Figure 5.4d	EELS	Weak	2.75 eV	3.92×10^4	1.25 eV
Figure 5.4d	EELS	Bulk	3.82 eV	2.19×10^4	0.34 eV
Figure 5.4d	EELS	SA	2.83 eV	4.53×10^4	0.70 eV
Figure 5.4d	EELS	Trans	3.58 eV	6.30×10^4	0.37 eV
Figure 5.5d	EELS	Weak	2.75 eV	3.92×10^4	1.25 eV
Figure 5.5d	EELS	Bulk	3.82 eV	2.19×10^4	0.34 eV
Parameters for power-law background fits			$f(\omega) = A \cdot \omega^k + c$		
Figure			A	k	c
Figure 5.1f			4.56×10^6	-0.81	-5.17×10^5
Figure 5.1g			1.57×10^6	-0.81	1.23×10^6
Figure 5.4d (Point 1)			1.03×10^5	-2.44	1.85×10^5
Figure 5.4d (Point 2)			6.30×10^4	-1.14	9.97×10^4
Figure 5.5d			1.23×10^5	-1.10	1.70×10^5

Below are all the parameters for all of the fits across both spectroscopies. All plasmon peaks are fitted with Lorentzians. The CL peaks are fitted in wavelength and then converted to eV before they are plotted in Chap. 5. The background corresponding to the ZLP is fit with a power law. LA, OOP, and SA refer to the dominant longitudinal modes, "Trans" is the transverse plasmon, and "Bulk" is the bulk plasmon referenced throughout Chap. 5. "Weak" refers to the weak plasmon tail discussed in Sect. 5.2.4

Appendix C
Sample Preparation for STEM Analysis

C.1 Solid-State Device Cross-Sections with Dual Beam FIB/SEM

In order to perform high-quality cross-sectional analysis of solid-state devices, a dual beam focused ion beam(FIB)/SEM is used to cut lamella out from a die. Figure C.1 shows an overview of the FIB extraction process. A die, with many different devices, is sent from IMEC (Fig. C.1a). Each device is localized between gate, source, and drain pins (or others depending on the nature of the device) utilized by our electrical engineering collaborators (Fig. C.1b). Once the devices are tested, the die is sent to us for cross-sectional extraction; the device manufacturers (or our collaborators) give a specific two-dimensional line desired for the cross section (Fig. C.1c). Finally, the cross section is milled and extracted and can be used for STEM analysis (Fig. C.1d). The difficult portion is getting from Fig. C.1c, d. Here, I provide the recipe I used for preparing the samples.

Figure C.2 shows the process used in the dual beam FIB for sample extraction. The FIB uses a focused Ga+ beam with a variable beam current (accelerating voltage is kept at 30 kV for this recipe, but it can be lowered as well) to both mill away parts of the sample as needed, as well as to provide an source for thin film deposition. Figure C.2a starts off with the schematic from Fig. C.1c, for the region of interest (ROI) in the sample. Once that region is found, a platinum gas is injected into the system near the film, and the FIB probe (operated at a relatively low current, either 50 or 100 pA) binds the platinum to the surface of the sample above the desired region (Fig. C.2b). This is done so that when the beam is brought up to its higher milling current the ROI is undamaged.

Once the ROI is protected, the cross-sectional lamella can be formed (Fig. C.2c). This is done by milling out large trapezoids out from the surrounding region, creating a thin wall (or lamella) that contains the region of interest. The lamella at this point is only connected to the bulk sample at the right side and the bottom.

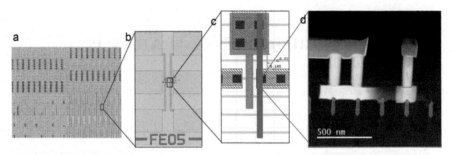

Fig. C.1 Overview of FIB lamella extraction. (**a**) Die shipped from manufacturer. (**b**) Individual device tested by electrical engineering group. (**c**) Requested cross section is given on schematic. (**d**) Cross section removed from that area and imaged in STEM

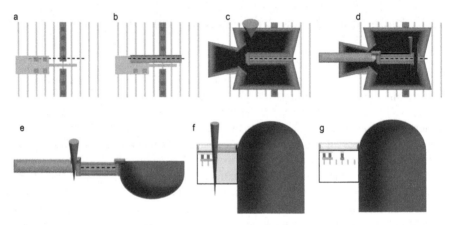

Fig. C.2 Cross-sectional lamella extraction process. The process used to extract the cross-sectional lamella from a die for STEM analysis. A full description is included in the main text

Now, I use a high beam current; 2 nA is a safe general value, but up to 8 nA have been used successfully.

To extract the lamella, a micromanipulator is brought in and abutted to the side of the sample (Fig. C.2d). A small platinum patch is deposited between the manipulator and the lamella to weld the two together. Once the weld is in place, the FIB beam cuts off the two remaining connectors to the die, at the bottom and at the side, leaving the lamella connected to only the micromanipulator.

Next, the micromanipulator is brought to the TEM grid (by moving the grid and not the micromanipulator, as the weld between the manipulator and the lamella is extremely fragile and can break easily. The lamella is then pressed up against the side of the TEM grid, and another platinum patch welds lamella to the sidewall of the TEM grid (Fig. C.2e). Then the sample FIB cuts away the weld between the micromanipulator and the lamella, leaving the lamella attached purely to the TEM grid. Once attached to the TEM grid, the lamella is highly stable. This is partly due to the fact that the TEM grid is much more stable than the micromanipulator, so vibrations are damped out before they strain the weld, and partly because the entirety of the lamella can be welded to the sidewall of the TEM, while only the tip of the lamella can be welded to the manipulator.

Finally, the sample is thinned to TEM thicknesses. This is done by rotating the TEM grid so that the FIB beam is parallel to the side of the lamella. Then gentle currents (50–100 pA again) are used to mill away the side of the lamella layer by layer until the ROI is revealed in cross section (Fig. C.2f). Once the entire ROI has been thinned down to electron transparency, the sample is ready for analysis (Fig. C.2g).

On a closing remark, the method used for the samples in this involved leaving the sample slightly thicker than ideal (~120–150 nm thick) and then using an argon ion polisher at extremely low voltages (900 V) to remove the final ~70 nm of lamella. The argon ion polisher is much more gentle than the gallium ion beam in the FIB and results in less damage to the sample. However, others have had good results by reducing the FIB beam to 10 kV and using a 30 pA probe to perform the final gentle polishing, in the absence of a nanomill.

C.2 Direct Sample Preparation of Nanospiral Arrays with EBL

The nanospiral arrays used in this dissertation are fabricated with a multi-step process that results with the nanospirals being written directly onto SiN windows. Figure C.3 shows a schematic of the process from beginning to end. The first step is to generate a computer-aided design (CAD) file with the desired per-chip array on. Figure C.3a shows an example of an array; note that this is just representative of the array and is not to scale; in the actual preparation, different arrays are used to suit different purposes depending on the aspect of the nanospiral being studied. Additionally, the CAD file is set to create an array of the nanospiral arrays. The final CAD file writes 625 separate arrays in a 25×25 pattern, where each column are all identical (in case samples are destroyed or lost or repeat tests are required), and each row is shot with a different dose to prevent having to do extensive dose testing beforehand. The spacing of the array is exactly 2.59 mm in both the x and y direction; this is done so that the corner-to-corner size of the final sample is 3 mm and will fit in STEM sample holders. The CAD file is then loaded into the EBL computer, which writes the design onto the Si wafer.

The wafer is 300 μm thick and 4" in diameter with an outer layer of SiN on both sides to help with the final etching stage. For the data shown in this dissertation, the arrays were made on 50-nm-thick SiN films; however, new iterations have successfully produced samples 25-nm-thick films, and 15-nm films are being attempted. Before the EBL, double-layer spin coating is performed using PMMA 495 for the first layer and PMMA 900 for the second layer; each layer is spun at 4000 rpm for 45 s and then baked at 180 °C for 5 min resulting in layers of 200 nm and 100 nm, respectively. The EBL writes the nanospirals, along with two large alignment markers, directly onto the spin-coated mask of the wafer. The wafer is then developed in a 1:3 mixture of MIBK and IPA and then submerged in acetone to remove the lithographic patterns from the mask (Fig. C.3b). The sample is now ready for metal deposition; for all samples, a thin 5-nm layer of Cr is deposited first to aid adhesion, and then the Au is laid down with DC sputtering from a thin gold foil (Fig. C.3c). Using a thin gold foil instead of a target allows for higher-quality films but requires a low power on the plasma; our experiments use 60 V at 7 W which results in a deposition rate of 1.92 nm/min; gold is deposited for different thicknesses in different samples. Once the gold is deposited, the sample is submerged in acetone and sonicated for 5 min for liftoff of the PMMA mask. After the liftoff, the result is gold nanospiral arrays in a 25 × 25 grid pattern (Fig. C.3d).

Now the nanospiral arrays are fully ready; however, they have still to be developed into usable STEM samples. For this stage photolithography is used. A single layer mask of S1818 is spin-coated onto the backside of the wafer at 3000 rpm for 45 s, resulting in a 40-nm-thick layer, and then baked at 115 °C for 1 min. For photolithography, a physical mask is used that is also a 25 × 25 array of grids, shown in Fig. C.3e. In this mask, the shape of each grid in the array is a square hole in the middle (500 × 500 μm) and four, 50-μm-thick, rectangular slits at the half way point between the square holes. The purpose of these slits is to provide a chipping line when the 25 × 25 array of samples is broken up into individual samples. The wafer is then loaded into a SUSS MA6 Mask Aligner. The mask is then back-aligned (using the alignment marks on the frontside of the wafer and on the photolithography mask as shown in Fig. C.3f) and exposed with a dose of 50 mW/s for 30 s. The photolithography mask is then developed in CD-26 for 2 min to prepare the wafer for reactive-ion etching (RIE) and rinsed in water to remove the mask from the exposed regions. After photolithography, the backside of the wafer is as shown in Fig. C.3g, with the thin SiN layer exposed in the pattern of the mask and the rest of the wafer covered by a ~400-nm layer of S1818; then it is exposed to a chlorine plasma for 1 min. The plasma completely removes the SiN in the exposed regions, but the masked regions are still masked; the entire wafer is then submerged in acetone for 5 min to remove the remaining S1818, leaving the SiN film showing again, but this time with a pattern in the shape of the mask etched out to reveal the Si beneath (Fig. C.3h).

The final stage of the preparation is a KOH etch. The sample is placed in a wafer holder that only exposes the backside of the wafer to surrounding liquids and then is immersed in KOH, heated to 80 °C, and left for 3 h. The KOH reacts with the SiN much more slowly the Si, so after 3 h the 300 μm of Si in the exposed regions has

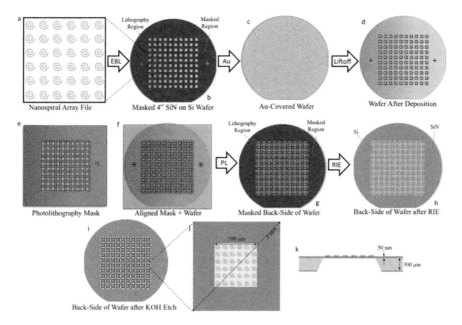

Fig. C.3 Nanospiral array preparation process. The process used to prepare wafers with 25 × 25 grids of STEM-ready samples with arrays of Au nanospirals. A full description is included in the main text

been dissolved, but the 50- nm-thick SiN layers on both sides of the wafer are still intact. The wafer is then removed from the KOH bath, rinsed, and air dried, leaving the nanospiral arrays on 50- nm-thick SiN windows on grids of 2.59×2.59 mm chips (3 mm corner to corner), ready to be broken up into individual samples and analyzed in the STEM (Fig. C.3i). The top view and side view of the finished product of an individual chip are shown in Fig. C.3j, k (not to scale).

Curriculum Vitae

FULL NAME: Jordan Adam Hachtel
DATE OF BIRTH: September 23rd 1985
INSTITUTION: Oak Ridge National Laboratory
Oak Ridge, Tennessee, USA
EMAIL: hachtelja@ornl.gov

Appointments Held

OCT 2016–*Current*	OAK RIDGE NATIONAL LABORATORY *Postdoctoral Research Associate*
JAN 2012–OCT 2016	VANDERBILT UNIVERSITY and OAK RIDGE NATIONAL LABORATORY *Graduate Research Assistant*
AUG 2011–MAY 2012	VANDERBILT UNIVERSITY *Graduate Teaching Assistant*
MAY 2006–MAY 2008	UNIVERSITY OF COLORADO AT BOULDER *Undergraduate Research Assistant*

Education

DEC 2016 — **Doctor of Philosophy** in PHYSICS, *Vanderbilt University*
Advisor: Prof. Sokrates T. Pantelides

MAY 2008 — **Bachelor of Arts** in PHYSICS, *University of Colorado at Boulder*
Bachelor of Science in APPLIED MATHEMATICS, *University of Colorado at Boulder*

Awards, Grants, and Honors

2017 | Springer Publishing Ph.D. Thesis Award

2015 | Microscopy & Analysis 2015 Presidential Scholar Award
Best Oral Presentation at 2015 TN-SCORE Annual Conference

2012–13 | Summer Research Grants awarded from Vanderbilt University

2008 | Dean's List for Spring Semester in College of Arts and Sciences at University of Colorado

2007–08 | Dean's List in College of Engineering at University of Colorado

2006–08 | Undergraduate Research Grants awarded from Undergraduate Research Opportunity Program within the National Science Foundation

Professional Activities

AUG 2015–*Current* | Member of the Materials Research Society

FEB 2013–*Current* | Member of the American Physical Society

JAN 2012–MAY 2013 | Team Leader for Vanderbilt Students Volunteer for Science

AUG 2004–DEC 2007 | Member of the Society of Physics Students at University of Colorado

Advisors and Collaborators

Oak Ridge National Laboratory	**Dr. Juan Carlos Idrobo** (Postdoctoral Advisor), Dr. Matthew Chisholm, Dr. Andrew Lupini, Dr. Anas Mouti, Dr. Ritesh Sachan, Dr. Benjamin Lawrie, Dr. Raphael Pooser, Dr. Philip Evans, Dr. Eugene Dumitrescu, Dr. Rohan Mishra, Dr. Ivan Kravchenko, Dr. Scott Retterer, Dr. Roderick Davidson
Vanderbilt University	**Prof. Sokrates Pantelides** (Ph.D. Advisor), *Prof. Richard Haglund* (Ph.D. Committee Member), *Prof. Kalman Varga* (Ph.D. Committee Member), Prof. Ronald Schrimpf, Prof. Daniel Fleetwood, Prof. Robert Reed, Prof. Norman Tolk, Dr. Yevgeniy Puzyrev, Prof. Joshua Caldwell, Prof. Xiao Shen, Dr. En Xia Zhang, Dr. Cher Xuan, Zhang, Dr. Guo Xing Duan, Dr. Liang Wang, Dr. Daniel Mayo, Dr. Kai Ni, Ms. Claire Marvinney, Mr. Oleg Ovchinnikov, Mr. Matthew Feldman
University of Tennesseeat Knoxville	*Prof. Gerd Duscher* (Ph.D. Committee Member), Prof. Ramki Kalyanaraman, Prof. Phillip Rack, Dr. Yueying Wu, Dr. Ondrej Dyck, Dr. Abhinav Malasi, Ms. Shaofang Fu
University of Colorado at Boulder	**Prof. James Smith** (Undergraduate Advisor), **Prof. William Ford** (Undergraduate Advisor), Dr. Frederick Blanc, Dr. James Hirschauer, Dr. Keith Ulmer, Mr. Zachary Clifton, Mr. Joe Becker
Rice University	Prof. Pulickel Ajayan, Prof. Boris Yakobson, Dr. Chandra Tiwary, Dr. Vidya Kochat, Dr. Alex Kutana, Mr. Amey Apte, Ms. Sandhya Susarla
ICMAB-CSIC	Prof. Anna Roig, Prof. Jaume Gázquez, Dr. Anna Laromaine, Dr. Marti Gich, Dr. Siming Yu
IMEC	Dr. Jérôme Mitard, Dr. Dimitri Linten, Dr. Nadine Collaert, Dr. Jacopo Franco, Dr. Liesbeth Witers
University of Texas at Austin	Prof. Delia Milliron, Mr. Ankit Agrawal, Mr. Shin Hum Cho
National University of Singapore	*Prof. Stephen J. Pennycook* (Ph.D. Committee Member)

Peer-Reviewed Journal Articles

In Press	S. Susarla[†], A. Kutana[†], **J.A. Hachtel**[†], V. Kochat, A. Apte, R. Vajtai, J.C. Idrobo, B.I. Yakobson, C.S. Tiwary, P.M. Ajayan **Quaternary two-dimensional (2D) transition metal dichalcogenides (TMDs) with tunable bandgap.** *Adv. Mater.*
	S. Susarla, V. Kochat, A. Kutana, **J.A. Hachtel**, J.C. Idrobo, R. Vajtai, B.I. Yakobson, C.S. Tiwary, P.M. Ajayan **Phase segregation behavior of 2D transition metal dichalcogenide binary alloys induced by dissimilar substitution.** *Chem. Mater.*
2017	E.X. Zhang, D.M. Fleewood, **J.A. Hachtel**, C. Liang, R.A. Reed, M.L. Alles, R.D. Schrimpf, D. Linten, J. Mitard, M.F. Chisholm, S.T. Pantelides. **Total ionizing dose effects on Ge pMOS FinFETs on bulk Si.** *IEEE Trans. Nucl. Sci.* **64**, 226
2016	G.X. Duan, **J.A. Hachtel**, E.X. Zhang, C.X. Zhang, D.M. Fleetwood, R.A. Reed, D. Linten, J. Mitard, M.F. Chisholm, S.T. Pantelides. **Effects of negative-bias-temperature-instability on low frequency noise in SiGe pMOSFETs.** *IEEE Trans. Device Mater. Rel.* **16**, 541
	J.A. Hachtel, S. Yu, A.R. Lupini, S.T. Pantelides, M. Gich, A. Laromaine, A. Roig. **Gold nanotriangles decorated with superparamagnetic iron oxide nanoparticles: a compositional and microstructural study.** *Faraday Discussions* **191**, 215
	J.A. Hachtel, C. Marvinney, A. Mouti, D.C. Mayo, R. Mu, S.J. Pennycook, A.R. Lupini, M.F. Chisholm, R.F. Haglund, S.T. Pantelides. **Probing plasmons in three dimensions by combining complementary spectroscopies in an scanning transmission electron microscope.** *Nanotechnology* **27**, 155202
2015	L. Wang, E.X. Zhang, R.D. Schrimpf, G.X. Duan, **J.A. Hachtel**, D.M. Fleetwood, R.A. Reed, M.L. Alles, S.T. Pantelides, G. Bersuker, C.D. Young. **Total Ionizing Dose Effects on Ge Channel pFETs with Raised $Si_{0.55}Ge_{0.45}$ Source/Drain.** *IEEE Trans. Nucl. Sci.* **62**, 2412
	S. Yu, **J.A. Hachtel**, M.F. Chisholm, S.T. Pantelides, A. Laromaine, A. Roig. **Magnetic gold nanotriangles by microwave-assisted polyol synthesis.** *Nanoscale* **7**, 14039
	G.X. Duan, **J. Hachtel**, X. Shen, E.X. Zhang, C.X. Zhang, B.R. Tuttle, D.M. Fleetwood, R.D. Schrimpf, R.A. Reed, J. Franco, D. Linten, J. Mitard, L. Witters, N. Collaert, M.F. Chisholm, S.T. Pantelides. **Activation Energies for Oxide- and Interface-Trap Charge Generation Due to Negative-Bias Temperature Stress of Si-Capped SiGe-pMOSFETs.** *IEEE Trans. Device Mater. Rel.* **15**, 352
	J.A. Hachtel, R. Sachan, R. Mishra, S.T. Pantelides. **Quantitative first-principles theory of interface absorption effects in multilayer heterostructures.** *App. Phys. Lett.* **107**, 091908
2014	G.X. Duan, C.X. Zhang, E.X. Zhang, **J. Hachtel**, D.M. Fleetwood, R.D. Scrimpf, R.A. Reed, M.L. Alles, S.T. Pantelides, G. Bersuker, C.D. Young. **Bias Dependence of Total Ionizing Dose Effects in SiGe-SiO_2/HfO_2 MOS FinFETs.** *IEEE Trans. Nucl. Sci.* **61**, 2834

[†]Equal Contributor

Peer-Reviewed Journal Articles continued...

2014 | A.J. Bevan et al., **The Physics of B Factories.** *Eur. Phys. J. C* **74**, 1

2013 | B. Aubert et al., **The BaBar detector: Upgrades, operation, and performance.** *Nuclear Instruments and Methods in Physics Research A: Accelerators, Spectrometers, Detectors, and Associated Equipment* **729**, 615

2008 | B. Aubert et al., **Observation of $\eta\rho^+$ and search for B^0 decays to $\eta'\eta$, $\eta\pi^0$, $\eta'\pi^0$, and $\omega\pi^0$.** *Phys. Rev. D* **78**, 011107(R)

2007 | B. Aubert et al., **Search for Neutral B-Meson Decays to $a_0\pi$, $a_0 K$, $\eta\rho^0$, and ηf^0.** *Phys. Rev. D* **75, 111102(R)**

Peer-Reviewed Conference Proceedings

In Press | J.A. Hachtel, S.Y. Cho, R.B. Davidson II, M.F. Chisholm, J.C. Idrobo, R.F. Haglund, S.T. Pantelides, B.J. Lawrie **Observing Nanoscale Orbital Angular Momentum in Plasmon Vortices with Cathodoluminescence.** *Microscopy and Microanalysis 2017*

J.A. Hachtel, R.B. Davidson II, R.F. Haglund, S.T. Pantelides, B.J. Lawrie, J.C. Idrobo **Near-Field Mid-Infrared Plasmonics in Complex Nanostructures with Monochromated Electron Energy Loss Spectroscopy.** *Microscopy and Microanalysis 2017*

J.A. Hachtel, S. Susarla, V. Kochat, C.S. Tiwary, P.M. Ajayan, J.C. Idrobo **Directly Identifying Phase Segregation in 2D Quaternary Alloys.** *Microscopy and Microanalysis 2017*

2016 | J.A. Hachtel, R.B. Davidson II, M.F. Chisholm, B.J. Lawrie, R.F. Haglund, S.T. Pantelides **Unveiling Complex Plasmonic Resonances in Archimedean Nanospirals Through Cathodoluminescence in a Scanning Transmission Electron Microscope.** *Microscopy and Microanalysis* **22 (Suppl. 3)**, 266

2015 | J.A. Hachtel, A. Mouti, D.C. Mayo, C.E. Marvinney, R. Mu, S.J. Pennycook, A.R. Lupini, M.F. Chisholm, R.F. Haglund, S.T. Pantelides. **Probing Plasmons in Three Dimensions Within Random Morphology Nanostructures.** *Microscopy and Microanalysis* **21 (Suppl. 3)**, 1683

J.A. Hachtel, D.C. Mayo, C.E. Marvinney, A. Mouti, R. Mu, S.J. Pennycook, A.R. Lupini, M.F. Chisholm, R.F. Haglund, S.T. Pantelides. **Direct Observation of Plasmonic Enhancement of Emission in Ag-Nanoparticle-Decorated ZnO Nanostructures.** *Microscopy and Microanalysis* **21 (Suppl. 3)**, 2389

J.A. Hachtel, D.C. Mayo, A. Mouti, C.E. Marvinney, R. Mu, S.J. Pennycook, A.R. Lupini, M.F. Chisholm, R.F. Haglund, S.T. Pantelides. **Spatially-Resolved, Three-Dimensional Investigation of Surface Plasmon Resonances in Complex Nanostructures.** *CLEO: 2015 OSA Technical Digest*, paper FTh1E.S

Journal Articles Under Review

- **J.A. Hachtel**, R.B. Davidson II, A.R. Lupini, B.J. Lawrie, R.F. Haglund, S.T. Pantelides **Near-field polarization selectivity of complex modes in plasmonic Archimedean nanospirals through cathodoluminescence imaging**. *ACS Photon.*
- **J.A. Hachtel**, S.Y. Cho, R.B. Davidson II, M.F. Chisholm, R.F. Haglund, J.C. Idrobo, S.T. Pantelides, B.J. Lawrie **Spatially and spectrally resolved orbital angular momentum interactions in plasmonic vortex generators**. *Light Sci. Appl.*
- V. Kochat, A. Apte, **J.A. Hachtel**, H. Kumazoe, A. Krishnamoorthy, S. Susarla, J.C. Idrobo, F. Shimojo, P. Vashishta, R. Kalia, A. Nakano, C.S. Tiwary, P.M. Ajayan **Re-doping in transition metal dichalcogenides as a new route to tailor structural phases and induced magnetism**. *Adv. Mater.*
- Y. Puzyrev, X. Shen, C.X. Zhang, **J.A. Hachtel**, K. Ni, B. Choi, O. Ovchinnikov, R.D. Schrimpf, D.M. Fleetwood, S.T. Pantelides, E.X. Zhang. **Memristive devices from ZnO nanowire bundles and meshes**. *App. Phys. Lett.*

Invited Seminars

2017	**J.A. Hachtel. Advanced electron microscopy techniques for nanoscale analysis of complex optical phenomena**. *University of Memphis: Department of Physics*. Memphis, Tennessee, USA. January 27th
2016	**J.A. Hachtel. Understanding the Nanoscale Optical Response of Complex Structures**. *Oak Ridge National Laboratory: Center for Nanophase Materials Sciences*. Oak Ridge, Tennessee, USA. June 14th

Invited Oral Presentations

2015	**J.A. Hachtel**, C.E. Marvinney, R. Mishra, A. Mouti, D.C. Mayo, R. Mu, S.J. Pennycook, A.R. Lupini, M.F. Chisholm, R.F. Haglund, S.T. Pantelides. **Understanding the Nanoscale Response of Complex Optical Structures**. *16th Annual Nanoscience and Nanotechnology Forum*. Nashville, Tennessee, USA. October 14th
2014	**J.A. Hachtel**, A. Mouti, D.C. Mayo, C.E. Marvinney, R. Mu, S.J. Pennycook, M.F. Chisholm, R.F. Haglund, S.T. Pantelides. **Probing emission and plasmons in nanostructures in STEM through combined spectroscopies**. *81st Annual SESAPS Meeting*. Columbia, South Carolina, USA. November 14th

Invited Poster Presentations

2015	**J.A. Hachtel**, C.E. Marvinney, A. Mouti, D.C. Mayo, R. Mu, S.J. Pennycook, A.R. Lupini, M.F. Chisholm, R.F. Haglund, S.T. Pantelides. **Probing Plasmons in Three Dimensions Through Complementary Spectroscopies**. *Microscopy & Microanalysis 2015 Meeting*. Portland, Oregon, USA. August 5th

Contributed Oral Presentations

2017   **J.A. Hachtel**, S.Y. Cho, R.B. Davidson II, M.F. Chisholm, R.F. Haglund, S.T. Pantelides, B.J. Lawrie, J.C. Idrobo **Near-Field Detection and Application of Optical Orbital Angular Momentum in the Electron Microscope**. *2017 MRS Spring Meeting & Exhibit*. Phoenix, AZ, USA. April 21st

J.A. Hachtel Finding, Labelling, and Analyzing Atoms Simply with Python. *Nion Swift Workshop III*. Bad Mittendorf, Austria. March 10th

2016   **J.A. Hachtel**, R.B. Davidson II, M.F. Chisholm, B.J. Lawrie, R.F. Haglund, S.T. Pantelides **Unveiling Complex Plasmonic Resonances in Archimedean Nanospirals Through Cathodoluminescence in a Scanning Transmission Electron Microscope**. *Microscopy & Microanalysis Meeting 2016*. Columbus, Ohio, USA. July 26th

J.A. Hachtel, R.B. Davidson II, A.R. Lupini, B.J. Lawrie, R.F. Haglund, S.T. Pantelides. **Complex Near-Field Plasmonic Response of Au Nanospirals**. *APS March Meeting 2016*. Baltimore, Maryland, USA. March 14th

2015   **J.A. Hachtel**, C.E. Marvinney, A. Mouti, D.C. Mayo, R. Mu, S.J. Pennycook, A.R. Lupini, M.F. Chisholm, R.F. Haglund, S.T. Pantelides. **Probing plasmons in three-dimensions through combined spectroscopies in the electron microscope**. *2015 MRS Fall Meeting & Exhibit*. Boston, Massachusetts, USA. December 2nd

J.A. Hachtel, D.C. Mayo, C.E. Marvinney, R. Mu, S.J. Pennycook, M.F. Chisholm, R.F. Haglund, S.T. Pantelides. **Direct Observation of Plasmonic Enhancement of Emission in Ag-Nanoparticle-Decorated ZnO Nanostructures**. *Microscopy & Microanalysis 2015 Meeting*. Portland, Oregon, USA. August 6th

J.A. Hachtel, C.E. Marvinney, R. Mishra, A. Mouti, D.C. Mayo, R. Mu, S.J. Pennycook, A.R. Lupini, M.F. Chisholm, R.F. Haglund, S.T. Pantelides. **Understanding Optical Phenomena at the Nanoscale**. *2015 TN-SCORE Annual Conference*. Nashville, Tennessee, USA. June 19th

J.A. Hachtel, A. Mouti, D.C. Mayo, C.E. Marvinney, R. Mu, S.J. Pennycook, M.F. Chisholm, R.F. Haglund, S.T. Pantelides. **Spatially-Resolved, Three-Dimensional Investigation of Surface Plasmon Resonances in Complex Nanostructures**. *CLEO: 2015*. San Jose, California, USA. May 14th

J.A. Hachtel, A. Mouti, D.C. Mayo, C.E. Marvinney, R. Mu, S.J. Pennycook, M.F. Chisholm, R.F. Haglund, S.T. Pantelides. **Probing plasmons in three dimensions in a scanning transmission electron microscope**. *APS March Meeting 2015*. San Antonio, Texas, USA. March 5th

2014   **J.A. Hachtel**, D.C. Mayo, A. Mouti, C.E. Marvinney, R. Mu, S.J. Pennycook, M.F. Chisholm, R.F. Haglund, S.T. Pantelides. **Cathodoluminescence Imaging of Plasmonic Resonances in Ag-Coated ZnO/MgO Core-Shell Nanowires in an Aberration-Corrected Scanning Transmission Electron Microscope**. *8th International Workshop on Zinc Oxide and Related Materials*. Niagara Falls, Ontario, Canada. September 10th

Contributed Oral Presentations continued...

2014 | **J.A. Hachtel**, R. Mishra, S.J. Pennycook, S.T. Pantelides. **Infrared absorption enhancement at nickel-silicide/silicon interfaces**. *APS March Meeting 2014*. Denver, Colorado, USA. March 5th

2013 | **J.A. Hachtel**, R. Sachan, O. Dyck, S. Fu, X. Shen, C. Gonzalez, P.D. Rack, G. Duscher, R. Kalyanaraman, S.T. Pantelides. **Absorption enhancement in amorphous silicon thin films via plasmonic resonances in nickel silicide nanoparticles**. *APS March Meeting 2013*. Baltimore Maryland, USA. March 21st

Contributed Poster Presentations

2017 | **J.A. Hachtel**, S.Y. Cho, R.B. Davidson II, R.F. Haglund, S.T. Pantelides, M.F. Chisholm, B.J. Lawrie, N.V. Lavrik, J.C. Idrobo. **Orbital Angular Momentum in the Electron Microscope**. *8th International Workshop on Electron Energy Loss Spectroscopy and Related Techniques*. Okuma, Okinawa, Japan. May 16thth

2015 | **J.A. Hachtel**, C.E. Marvinney, R. Mishra, A. Mouti, D.C. Mayo, R. Mu, S.J. Pennycook, A.R. Lupini, M.F. Chisholm, R.F. Haglund, S.T. Pantelides. **Understanding Optical Phenomena at the Nanoscale**. *2015 TN-SCORE Annual Conference*. Nashville, Tennessee, USA. June 18th

2014 | **J.A. Hachtel**, R. Mishra, S.T. Pantelides. **Atomistic Optical Absorbance in Complex Heterostructures** *2014 TN-SCORE Annual Conference*. Nashville, Tennessee, USA. June 26th

2012 | **J.A. Hachtel**, S. Fu, R. Sachan, O. Dyck, C. Gonzalez, P.D. Rack, G. Duscher, R. Kalyanaraman, S.T. Pantelides. **Absorption enhancement in amorphous silicon thin films via plasmonic resonances in nickel silicide nanoparticles**. *2012 TN-SCORE Annual Conference*. Nashville, Tennessee, USA. March 21st

Co-authored Oral Presentations

2017 | J. Hachtel, R. Davidson, M. Chisholm, R.F. Haglund, S. Pantelides, S.Y. Cho, <u>B. Lawrie</u>. **Nanochirality detection with vortex plasmon modes**. *CLEO: 2017*. Tirol, Austria. May 14th

<u>B. Lawrie</u>, R. Pooser, J. Hachtel, R. Davidson. **Photon bunching and antibunching in cathodoluminescence at high currents**. *6th International Topical Meeting on Nanophotonics and Metamaterials*. Tirol, Austria. January 5th

2016 | <u>E.X. Zhang</u> D.M. Fleetwood, **J.A. Hachtel**, C. Liang, R.A. Reed, M.L. Alles, R.D. Schrimpf, D. Linten, J. Mitard, S. Pantelides. **Total ionizing dose effects on Ge bulk *p*-MOS FinFETs**. *2016 National Space Radiation Effects Conference*. Portland, Oregon, USA. July 12th

<u>Y. Puzyrev</u>, X. Shen, K. Ni, C.X. Zhang, **J.A. Hachtel**, B. Choi, M.F. Chisholm, D.M. Fleetwood, R.D. Schrimpf, S.T. Pantelides. **Memristive switching of ZnO nanorod mesh**. *APS March Meeting 2016*. Baltimore, Maryland, USA. March 14th

Presenting Author Underlined

Co-authored Oral Presentations continued...

2015 | C.E. Marvinney, D.C. Mayo, **J.A. Hachtel**, J.R. McBride, W. Liu, H. Xu, Y. Liu, R. Mu, R.F. Haglund. **Modulated Photoluminescence in ZnO Core-Shell Nanowires with Plasmonic Nanoparticles**. *2015 MRS Spring Meeting & Exhibit*. San Francisco, California, USA. April 8th

2014 | G.X. Duan, C.X. Zhang, E.X. Zhang, **J.A. Hachtel**, D.M. Fleetwood, R.D. Schrimpf, R.A. Reed, M.L. Alles, S.T. Pantelides, G. Bersuker, C.D. Young. **Bias Dependence of Total Ionizing Dose Effects in SiGe-SiO$_2$/HfO$_2$ pMOS FinFETs**. *2014 National Space Radiation Effects Conference*. Paris, France. July 17th

Presenting Author Underlined

Co-authored Poster Presentations

2017 | B. Lawrie, S.Y. Cho, R. Davidson, **J. Hachtel**. **Nanochirality detection with vortex plasmon modes**. *6th International Topical Meeting on Nanophotonics and Metamaterials*. Tirol, Austria. January 6th

2015 | L. Wang, C.X. Zhang, G.X. Duan, R.D. Schrimpf, D.M. Fleetwood, R.A. Reed, I.K. Samsel, **J.A. Hachtel**, M.L. Alles, L. Witters, N. Collaert, D. Linten, J. Mitard, S. Pantelides, K.F. Galloway. **Total Ionizing Dose Effects on Ge Channel pFETs with Raised Si$_{0.55}$Ge$_{0.45}$ Source/Drain**. *2015 National Space Radiation Effects Conference*. Boston, Massachusetts, USA. July 16th

2014 | D.C. Mayo, C.E. Marvinney, **J.A. Hachtel**, M.F. Chisholm, S.T. Pantelides, R. Mu, R.F. Haglund. **Cavity-mode effects on exciton-plasmon coupling in ZnO/MgO core-shell nanowires decorated by Ag nanoparticles**. *8th International Workshop on Zinc Oxide and Related Materials*. Niagara Falls, Ontario, Canada. September 12th

Presenting Author Underlined

Updated on July 17th 2017

CPSIA information can be obtained
at www.ICGtesting.com
Printed in the USA
LVOW05*1823070118
562135LV00001B/3/P

9 783319 702582